MINDFUL SCEPTICS
GUIDE TO...

sustainability

Mark Dangerfield

Mindful

Published by Mindful Sceptics

Mindful Sceptics

Springwood, NSW 2777, Australia

mindfulsceptics.info

AI Use Statement

This book was conceived, written and edited by a human who occasionally used AI tools to assist with grammar, editing, summaries and focused searches.

Contents

Preface to the Series

Ever feel like you're drowning in information and controversy, unsure what to believe? You're not alone. In this age of information overload, finding clarity can seem like searching for a needle in a digital haystack.

But what if I told you that the key to navigating this chaos lies within you?

Welcome to the Mindful Sceptic Guides, your compass in the turbulent waters of modern discourse. These aren't your average self-help books or dry academic tomes. They're intellectual adventures designed to awaken your innate curiosity and sharpen your critical thinking skills.

Imagine being able to cut through the noise, to see past the spin and the hype. Picture yourself in conversations where your insights turn heads, where your words carry the weight of well-reasoned thought. That's the power of being a mindful sceptic, a power that's within your reach.

In the pages of *The Mindful Sceptic Guides*, you'll find a delightfully eclectic mix of crucial facts, thought-provoking ideas, and paradigm-shifting perspectives. We'll tackle big issues with a unique blend of scientific rigour and philosophical inquiry. From conservation to social justice, from food security to healthy diet, economic theory to personal growth, no topic is off-limits.

But here's the kicker. This isn't just about personal enlightenment. The stakes are higher than that. In a world grappling with complex, existential challenges, humanity's survival may depend on cultivating more mindful sceptics—people who can question dogma, seek evidence, and imagine new solutions to age-old problems.

Could you be one of these crucial thinkers? Could your ideas be the spark that ignites positive change?

There's only one way to find out. Dive into these guides with an open mind and a willingness to challenge your assumptions. Prepare for moments of revelation, ideas that will keep you up at night, and insights that will transform how you see the world and your place in it.

Are you ready to embark on this intellectual journey? To become a beacon of reason in a world that desperately needs it? Then, turn the page. Your adventure in mindful scepticism begins now.

Welcome to the Mindful Sceptic Guides. Let's change the world, one critical thought at a time.

Preface to this Guide

Human business-as-usual has delivered wealth beyond imagination for some, starvation for others, and a middling daily struggle to make ends meet for most of the world's 8 billion souls, their livestock and pets. But whatever you think about our modern world with its six-continent supply chains, super yachts, gas-guzzling SUVs, and 46% of the people living on less than $6.85 a day, on just about any measure, the human enterprise is an extraordinary business. And for the lucky ones like me, being a human is amazing.

I am a highly educated, middle-class white guy in his sixties who has been privileged to make a living as a research ecologist and educator over a career that has seen residence on three continents, visits to dozens of countries and an unrecorded number of habitats from savannas to rainforests. I've been about a bit and see the good, bad and a little of the ugly of what business as usual has done to people and nature. Most of the time I am fighting deep unease with the widening gap between our stated ambitions and the biophysical realities of a finite planet.

Researching tsetse fly management in Africa, land systems in Australia, carbon credits in Asia, and running an ecological consultancy that services clients worldwide, I've come to believe that a radical re-evaluation of our most cherished assumptions about progress, growth, and the human relationship to nature is essential. Only it's hard. It requires a willingness to ask awkward questions, challenge entrenched paradigms, and confront uncomfortable truths about our current trajectory's thermodynamic and ecological limits.

Rather than whinge or get stuck in a polemic, this book is an invitation to undertake critical self-examination, to peer behind the facade of business as usual and grapple with the profound implications of our ecological predicament. In the following chapters, we'll explore the fundamental contradictions of 'sustainable development,' the hidden costs of our fossil fuel addiction, the spectre of Jevons' paradox, and the harsh realities of feeding 8 billion people without unravelling the web of life.

But for all its unflinching analysis, this book is ultimately grounded in hope and possibility. I believe we can chart a path toward regeneration by reconnecting with ecological processes and reimagining our role as responsible stewards rather than reckless exploiters. Doing so will require humility, adaptability, and a willingness to think in longer timescales than the next quarterly report or election cycle. But the alternative is too grim to contemplate.

My goal in writing this book is not to provide a step-by-step blueprint or peddle the illusion of easy fixes but to spur a long-overdue conversation about the true dimensions of the sustainability challenge. If the ideas presented here provoke reflection, debate, and a renewed sense of urgency and commitment, I will consider it a success. Ironically, trolls would be fine, too.

The road ahead will not be easy, but, like the late Hans Rosling, I am a possibilist. The destination is a world where humanity lives in harmony with the natural systems that sustain us, and it is still within reach. Getting there will require our collective wisdom, creativity, and determination.

This book is for the curious and the concerned, the skeptics and the dreamers. It's for anyone who's ever looked at the world and thought, *There's got to be a better way*. My goal isn't to hand you a magic solution—those don't exist in the real world. Instead, I want to spark a conversation, light a fire under our collective backsides, and make you see sustainability not as a buzzword but as the challenge of our lifetime.

Are you ready to become a mindful sceptic? To question everything you thought you knew about progress, growth, and our place in nature? Then, turn the page. Our journey towards true sustainability begins now. Only buckle up; it's going to be one hell of a bumpy road.

Chapter One

Protect the bottom line

Our hard-wired inability to look past immediate gratification, coupled with corporations' single-minded pursuit of profits, poses an existential threat. Bullshit all you like; the bottom line is singular

Daydreamers die of thirst

Imagine for a moment that you are in the mind and physically fit body of your ancestor, who lived somewhere on the African savanna 70,000 years ago. The place where you live looks something like this...

Zebra graze in the Masai Mara, Kenya.

Remember, you were born and brought up in this place, so you know the view and are accustomed to the sights. What happens next? You can't reach for your smartphone to take a selfie with the zebra because you don't have one. Your priorities are finding food, avoiding becoming food and getting to the nearest fresh water at the base of the hill in the distance before dark. There is no chance of bringing down any healthy zebra out in the open, but a good chance there is a lion or six between you and the water.

Your brain is working overtime at the signs, especially the sounds and smells of the savanna, to survive and glean what you can from the daylight hours. You will think about your fellows, the tribe, or the extended family you rely on to help with all the vigilance and foraging. Some thoughts of communication, perhaps even some empathy, allow you to act on keeping the bonds of cooperation tight.

You must pay attention to the moment, to the immediate needs, and to the risks involved in meeting them. There is little time or energy for musing on the future. Daydreamers die of thirst.

Fast forward to today, and your brain retains the adaptations evolved to survive without modern technology on the savanna. We worry about the present disproportionately more than the future for evolutionary reasons—short life spans, risk of food expropriation, and unpredictable environments. As Daniel Kahneman describes, our brain thinks fast, only occasionally slow, and cares much more about the present than the future.

Economists recognise this innate response and evaluate it through a 'discount rate' that makes the future disproportionately less costly than the present. The steeper the discount rate, the more 'addicted to the present' the person is. For example, drug and alcohol users, risk takers, persons with low I.Q. scores, people with severe cognitive workloads, and men, compared to women, tend to discount future events or concerns more sharply.

Here is a trivial ramification of our immediacy. When asked to select chocolate or fruit as a snack for the following Monday, individuals chose fruit 75% of the time. But 70% of people prefer chocolate as today's snack—we discount the future. Regarding movies next week, 63% prefer informative documentaries, while 66% prefer comedies or sci-fi for their viewing pleasure tonight—we discount the future and lap up the good stuff now.

The problem with our immediacy is that, for most people in Western societies, the evolutionary reasons for worrying about now no longer apply. We don't have to worry about food, water, or shelter because we invented economies and social systems that, for most people, take care of these things for today. But those same systems push the cost into a future that we have an innate ability to ignore. The fossil fuel pulse has generated wealth and a lot of people with the bandwidth to think up clever ways of keeping it all going. Instead, we spend our time worrying about the modern equivalent of finding food, avoiding becoming food and getting to the nearest fresh water. If you want to know what this worry looks like, watch an episode of The Real Housewives. Any city will suffice.

What the market needs

"*Mangwanani*,' whispered a wizen old man with two-tone hands held out in the polite way to greet a stranger.

"*Mangwanani, marara sei*?" I replied sheepishly because I was a white man in Africa with no right to be there.

"Some meilie for the picinins?" The old man asked, well aware I had reached the limit of my Shona.

"Here is some," I said, passing a 5kg bag of ground maize into the man's palms.

"*Ndatenda zvikuru*," the old man said with an impossibly wide smile.

"*Ndatenda*," I replied, loading the rest of my weekly groceries into the beaten-up Renault I had borrowed from a friend.

Such exchanges often happen between people all over the world, not always with the bag of maize meal but with a poor person forced to ask for food from a wealthy one.

I lived in Zimbabwe for two years in the late 1980s, when the country was less than a decade from the euphoria of independence. The early signs of economic turmoil were present, such as a shortage of foreign exchange and disruption to food production and processing. Still, in the supermarkets, all the basic foodstuffs were readily available. You could buy bread, milk, cheese, eggs, seasonal fruit and vegetables, and maize flour, the staple carbohydrate for most Zimbabweans. There was plenty of beef, pork and even the scrawny African chicken—all the foods needed to combine into a healthy diet, finances permitting.

At least 65 million people in Africa are impacted by acute food insecurity.

The fact that there weren't thirty varieties of breakfast cereal on the shelves didn't matter. Everyone was quite used to what was available, and most city dwellers with access to jobs were well-nourished. Much later, I saw a similar thing in Papua New Guinea. Supermarket shelves were well stocked, and people could buy food, but the variety of processed food was an order of magnitude less than a typical Western supermarket.

In my local Australian grocery store, I can buy bread, milk, cheese, eggs and so on, but I have to choose between multiple versions of each. And then there are thousands of processed foods to get my head around. A few raw materials are processed and then ultra-processed into a wide range of marketed and sold products, allowing multiple players into the marketplace. A typical modern supermarket carries 30,000 products, a mind-bending diversification of foods that comes with a so-called mature economy.

Consumers don't need ten varieties of eggs or 12 types of gluten-free biscuits, but the market has created variety anyway. The market thrives on trades, buyers and sellers, and it can accommodate any number of goods and services with actual or potential buyers. The only limitation is the capacity to exchange.

Markets are powerful. Regulated or not, they present an insatiable capacity for 'more making' let loose by the subsidy of cheap energy. Somewhat ironically, the success of the market mechanism in creating wealth has locked in growth; the system collapses without it. One of the many consequences is a plethora of goods and services that are not needed for human survival or well-being. They appear because they can be exchanged. Even if society decided that exponential growth wasn't possible because it isn't on a finite planet and chose to focus on essential goods and services, the market wouldn't listen. It cannot. Insatiable, maybe, but needs are deaf, dumb and blind.

Even if he asks politely, the market has no time for the old man without power to purchase.

The bottom line is singular

Robert Reich, the American economist who was US Secretary of Labor in the Clinton administration, says this about the pivotal players in markets.

> *CEOs won't do anything that hurts their bottom lines. They're in the business of making as much money as possible, not solving social problems. In fact, real social change would prevent them from doing many of the hugely profitable things they now do. Which means they won't change their ways unless they're required by law to change (and even then, only when the penalty times the probability of getting caught is higher than the profits from continuing anyway). Their soothing promises of social responsibility are intended to forestall such laws.*
> Robert Reich, Why Corporate Social Responsibility Is BS

And I have to agree. The bottom line is singular despite all the rhetoric. John Elkington, in his book Green Swans, seems to think otherwise. He borrows the swan analogy blatantly from the excellent Nassim Taleb whilst managing not to understand what Taleb meant by his black swans—unpredictably rare yet inevitable. After outlining many of the disasters and absurdities that capitalism has imposed on everyone through unfettered resource use and waste, Elkington would have us believe that a shift in thinking among CEOs is still possible. He seems to adopt this notion because he thinks a handful of them might be thinking about it.

Seeing the glass as half-full can make anything look optimistic. However, to mix the metaphors, the frog in the saucepan of water thought everything was fine at the halfway point. Perhaps feeling the heat in 2018, John Elkington decided to make a 'management concept recall' of his 25-year-old accountability concept, the triple bottom line, just as companies sometimes must do with faulty products. He coined the term as a challenge for business leaders to rethink capitalism and *'stop focusing solely on profits and expand their focus to include improving the lives of people and the health of the planet'.* He had concluded that the triple bottom line had been reduced to a mere accounting tool, a way of balancing trade-offs instead of doing things differently. It has morphed into corporate social responsibility or CSR for many companies. Indeed, who could argue that accounting for social and environmental consequences of actions was anything if not responsible?

> *Corporate social responsibility (CSR) is a self-regulating business model that helps a company be socially accountable—to itself, its stakeholders, and the public. By practicing corporate social responsibility, also called corporate citizenship, companies can be conscious of the kind of impact they are having on all aspects of society, including economic, social, and environmental.*
>
> Investopedia

CSR is a self-regulating way of thinking. I think about it for a bit and then move on, perhaps to profiteering. However, because I spent time thinking about my responsibility, I am responsible. Rhetoric has always been easy.

Elkington would have us avoid the pungent waste of CSR by looking for Green Swans instead.

> *A Green Swan is a profound market shift, generally catalyzed by some combination of Black or Gray Swan challenges and changing paradigms, values, mind-sets, politics, policies, technologies, business models, and other key factors. A Green Swan delivers exponential progress in the form of economic, social, and environmental wealth creation. At worst, it achieves this outcome in two dimensions while holding the third steady. There may be a period of adjustment where one or more dimensions underperform, but the aim is an integrated breakthrough in all three dimensions.*
>
> John Elkington, Green Swans

The implied win-win-win of the triple bottom line becomes a delightful but naive option compared to the much more likely win-lose outcomes of resource use. The green swan is a miraculous, rare, and fanciful evolution of the dinosaur lineage.

Obviously, Reich favours government action, and Elkington suggests that corporations can somehow invent green birds. Nassim Taleb is the pragmatic one. With his Black Swans, he reminds everyone that you can talk up what you like and have governments intervene, but unpredictable disturbance is inevitable and natural, no matter how you operate in the world. Expecting the unexpected is how to prepare for the future.

My experience across a career straddling business, academia, government and NGOs is that almost everyone is in the immediate, merrily discounting the future. Companies especially are ill-prepared for the unexpected disruption, but they know we are suckers for a good news story, and they churn them out for the feel-good they generate. It can be TBL, CSR or absurdly coloured birds; the message is a distraction from the negative consequences of protecting the prime directive, profits—classic smoke and mirrors.

The worrying trend is that the market has taken over politics to the point that not even politicians know the difference between politics and policy, happily ignoring the latter. I wonder if any government is even capable of figuring out the law changes Robert Reich deems necessary. Yet, I have zero expectation that corporations will voluntarily rethink capitalism. Someone else must do it for them. This can't be the governments or the bureaucracy that supports them, for both have lost the plot. Academics are having a go, but they lack dress sense. The best person to figure this all out and return responsibility to the world is a mindful sceptic.

Key points

- Our brains evolved to prioritize immediate needs and discount future consequences, making us innately short-sighted and prone to overconsumption.

- Modern market systems, driven by the profit motive, enable and encourage overabundance of products beyond what we truly need for well-being.

- Corporate social responsibility initiatives like the "triple bottom line" are viewed as rhetoric rather than genuine efforts to rethink business practices.

- Meaningful change will likely require government regulation and intervention, rather than relying on voluntary corporate action, to overcome psychological biases and misaligned economic incentives.

In the next chapter

Demand growth and supply uncertainties related to phosphate rock reserves, food waste, and hunger are just some of them. Let's look at pork consumption in China.

Chapter Two

Food

Industrial efficiency has increased food production but at the cost of fragilities, resource stresses, ethical failings around hunger and waste, and placing humanity's entropy-defying activities on a collision course with ecological limits.

Pork consumption in China

Eric from Accrington is on his favourite food app and orders 72 and 43, please, sweet and sour pork and Special fried rice. Delicious. While the order from the local Chinese comes to Eric via Uber Eats, he dreams of the little delights of pork in the fried rice and those sickly sweet chunks of pork in their gooey, sugary coats of deep-fried loveliness. If only he knew what the sugars were doing to his mitochondria.

No self-respecting Chinese citizen would consume such tragedies offered by a British takeaway; they do eat a lot of pork. Pork has been a big part of Chinese cuisine as far back as 7,000 BC, and by the 1900s, at least 70% of animal calorie intake in China came from pork. Today, consumers in China devour roughly 54 million tons of pork annually, the highest total worldwide. No surprise given the size of the population, but the per capita consumption has also grown steadily since the early

1980s. When I was born in 1961, per capita pork consumption in China was roughly 7 kg per year, whereas by 2000, it was over 30 kg per annum. Despite an outbreak of African Swine Fever in China in 2018 and 2019 that killed over 130 million pigs, consumers still averaged over 30 kg per annum throughout the 2000s.

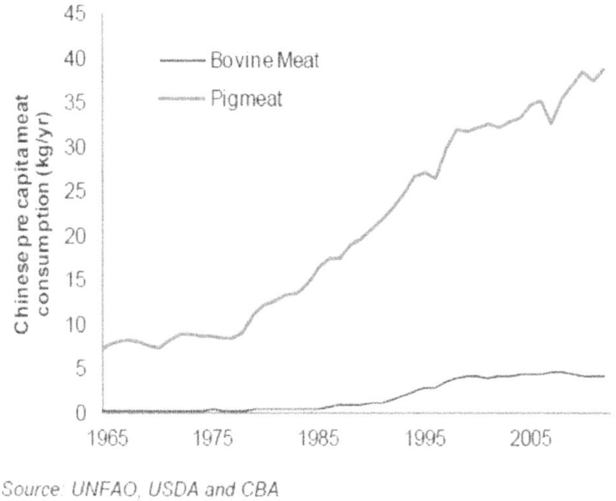

Pork consumption in China based on data from UN Food & Agriculture Organisation and USDA

Security of pork supply

Chinese companies have met the local demand for pork by increasing domestic production and acquiring foreign capacity. In 2013, the WH Group, formerly known as Shuanghui International Holdings, bought the American pork producer Smithfield for $4.7 billion, then the largest acquisition of a U.S. company by a Chinese business. Smithfield Foods boasts more than 40,000 U.S. employees and nearly 50 facilities across the U.S.

The purchase of overseas production capacity and maintenance of stockpiles has happened alongside growth in local pork production in China, which is increasingly intensive. Local producers use artificial intelligence to build capacity with two-thirds of China's pork production from large corporations using technological tools, including computational genetics and biotechnology, to breed pigs. Genetic control, automatic feeding, water dispensing systems, and strict exercise times mean pigs are farmed to a precise size, perfect for industrial farming. What the pigs eat has become a combination of genetically modified soybeans, grains, protein powders, and sometimes treated food waste. The food waste can contain pork—the added protein powders are often derived from pigs. In other words, the pigs are being fed to pigs.

This is a classic intensification of production. It simplifies the product, in this case, genetically modified pigs, and ensures that the food components are as cheap as possible. The scale and efficiency of these systems are undeniable. Pork products are more affordable than ever and sold to more consumers. What used to be a luxury is consumed by the majority. The market sees this as a virtuous cycle.

Historically, Chinese farmers raised hundreds of pig breeds of different sizes and attributes, and they adapted to local climates and diseases. Animals from these breeds could be fed leftovers and generate rich fertiliser for the fields. The regenerative, internal cycling of nutrients was vital to agricultural production in rural areas.

Small-scale and holistic production works well for smallholdings but fails to compete with the economies of scale from industrial processes. Ideal pigs reared to a uniform size in factory farms make for tough competition in urban markets, and that market is demanding.

Estimates are that 48 native pig breeds are reared in China

Production in 2020 was 41 million tonnes, down from 54 million tonnes in 2018 due to the African swine fever outbreak, but projected to return to this output. Intensification stabilises the market, increases profit and creates a near-perpetual demand curve. High production at low prices meets but also creates demand.

The challenge for China, which now has many citizens expected and expecting to buy pork, is to balance several factors affecting supply and demand, each with its values and advocates. Four of these are tough to balance together

- technology providing the capacity
- market pressures to produce food at low prices
- food security requirements of making sure that the prices don't become too volatile and cause instability
- the long-term capability of producing food under these conditions.

Technology can increase production at the cost of greater energy and nutrient inputs. Applied at scale this can keep prices low. Maintaining future production becomes less and less likely when this balance becomes unstable. Intensive production makes yield dependent on inputs and reduces resilience to disease outbreaks or supply chain disruption. Pork consumption in China has already been affected by both of these outcomes, with swine fever and COVID-19 supply chain disruption affecting supply. Production can be maintained if input dependencies hold but not if they fail. Be sure that the Chinese government knows all about this risk.

China is the only country with a pork reserve consisting of millions of live pigs and unaccountable tons of frozen pork, a reserve capacity from domestic and foreign sources. In 2008, when the country experienced food price rises, the government drew on these pork reserves to stabilise the price. Few governments want to risk destabilisation from the impact of food shortages and, when they can, will maintain food supply to stabilise food prices. The real reason for food reserves is to buffer against the effects of higher food prices that typically result in public discontent. Of course, pigs don't grow without the soy, grains and and protein powders from plants that have to be grown.

Peak phosphorus

The FAO report estimates that global fertiliser use in 2021 was 266 million tons, including ammonia, phosphoric acid, and potash combined. Although 266 million tons sounds like a lot, it is a meaningless number on its own. For some context, a typical medium-sized bulk cargo ship used to transport fertiliser displaces 50,000 tons, so we are talking about 5,320 shiploads of the stuff. That's a lot of ships.

In 24 hours, a large cargo ship can travel 1,000 km and consume 280,000 litres of fuel.

Most people think of nitrogen as essential for plant growth, and it is, but phosphorus is critical, too, especially in root crops. It is vital for cell division, especially for seedlings and young plants. Adding phosphorus to soil promotes root growth and winter hardiness, stimulates tillering, and can speed up plant maturity.

Phosphate rock minerals are the only significant global phosphorus resource used in agriculture. It used to be guano—accumulated bird and bat poo—full of nitrogen, phosphate and potassium. In the 19th century, the guano trade was pivotal in developing modern input-intensive farming. The high demand led the United States to pass the *Guano Islands Act* in 1856, which gave U.S. citizens who discovered a source of guano on an unclaimed island exclusive rights over the deposits. Have to love the arrogance. Once the German chemists Fritz Haber and Carl Bosch figured out an industrial method to artificially fix nitrogen in ammonium in the early 1900s, bird droppings became less popular. It was easy to process rock phosphate and add it to the ammonium as a critical ingredient in many inorganic fertilisers. Today, global agriculture relies on rock phosphate, with 220 million tons used in 2021.

By 1846, 462,057 tonnes of guano had been exported to Great Britain from Ichabod Island, off the coast of Namibia and surrounding islands.

Researchers estimated that the Earth's commercial and affordable phosphorus reserves could be depleted in 50–100 years, reaching peak phosphorus in approximately 2030. Others suggested that supplies would last for several hundreds of years. They also reported a lower global estimate of 71 billion metric tons of phosphate rock reserves, but this is still over 300 years of supply at the 2021 extraction rate. The International Fertilizer Development Center, IFDC, estimated worldwide phosphate reserves are around 60 billion metric tons of concentrate. In addition, the IFDC estimate of global phosphate rock resources is approximately 290 billion metric tons, which includes the unprocessed ore of the reserve estimates. Even though we would expect a bullish projection from a fertiliser development NGO, billions of tons of reserves should sound reassuring against millions of tons of annual use.

Where the phosphate rock comes from is as important as the volume. The United States Geological Survey reported global phosphate rock production in 2021 of 220 million metric tons, including 85 MMT from China and 38 MMT from Morocco. If China moved to choke supply, it would be catastrophic, as would a problem in Morocco, given that they make half the total volume between them. Such geographic concentration is a risk.

Morocco has occupied the disputed region of Western Sahara since 1975. If the reserves in this region are added to those in Morocco proper, they amount to 72% of all phosphate-rock reserves worldwide.

There is time for all manner of disasters to befall humanity before we run out of phosphate rock. The resource exists. The challenge is more about the accessibility of these reserves. How easy will it be to dig them out of the ground, and how costly? Not to mention the geopolitics of a resource crucial to feeding everyone.

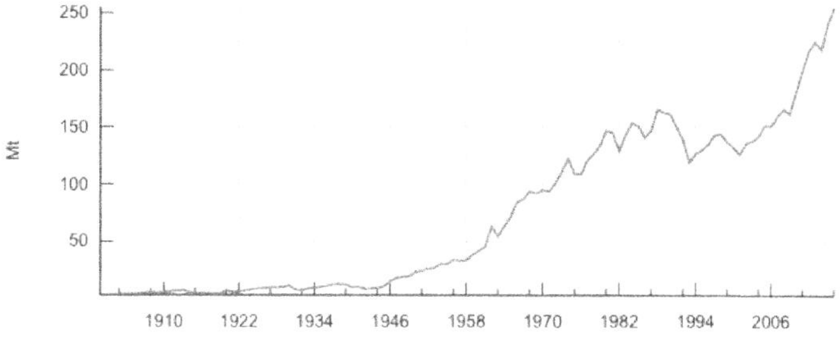

Graph by Stefan Pohl - Own work, CC0, https://commons.wikimedia.org/w/index.php?curid=38300993

Given the resource volume, logically, there is no peak phosphorus yet, but there is still a sovereign risk, not to mention the disruption should supply chain delays affect a proportion of those 5,000+ shiploads. Even when reserves are healthy, supply might not be. Clearly, there are all sorts of risks that come with a global market. Meanwhile, the demand is climbing as agricultural intensification spreads and depleted soils require more fertiliser than ever to maintain crop yields.

An embarrassment for the well-fed

I never remember going to bed hungry except for the self-imposed stomach pangs from my latest fad diet. The fact that I diet at all is a giveaway. I have access to too much food or choose to consume the wrong kinds. I am one lucky bugger. All I have to do is carry the food waste into the trash and, once a week, put the trash can where the truck can collect it from the sidewalk.

Enough food is produced worldwide to feed everyone well, thanks mostly to the agricultural revolutions that have absorbed fossil fuel inputs and channelled nature into agriculture. There is no reason from the supply side for anyone to be food insecure. Yet according to the World Food Program, at least 828 million people don't know where to get their next meal. Imagine that as one in every ten people you see on your morning commute. And some of them might be in your train carriage.

Here is the reality in 2021 as described by the United Nations in their online chronicle

> *Each day, 25,000 people, including more than 10,000 children, die from hunger and related causes. Some 854 million people worldwide are estimated to be undernourished, and high food prices may drive another 100 million into poverty and hunger. The risks are particularly acute among those who must spend at least 60 per cent of their income on food: the urban poor and displaced populations, the rural landless, pastoralists and the majority of smallholder farmers.*
>
> John Holmes, United Nations Under-Secretary-General for Humanitarian Affairs

I am lucky because I live in Australia, where the average household spends 10% of its income on food. This means that food prices can double, and whilst I may complain, it is well within my means to afford the extra cost. But what happens if food prices double and your current spending on food is 60% of your income? There is not enough money to spend on food, let alone anything else. This is the caricature of food insecurity—poor people living hand to mouth in weak economies.

I can remember back to the 1980s and Live Aid with famous pop stars earnestly singing *Feed the World* with its horrendously patronising neo-colonial lyrics like *"There's a world outside your window. And it's a world of dread and fear"* and "*Do they know it's Christmas time at all?*" In other words, hunger happens to other people who are probably black or brown and live far away. But the messaging worked. Live Aid raised $127 million for famine relief, equivalent to $350 million in 2022, and earned Bob Geldof a knighthood.

It is very different now. Not all the food-insecure people live in remote corners of the world. Today, they exist in every country. According to the USDA, 34 million people in the United States are food insecure—one in every ten people, the global average. It gets worse. One in six people, 53 million, turned to food programs in 2021. In 2018, the Supplementary Nutritional Assistance Program (SNAP) supported roughly 40 million Americans, with 9.2% of American households receiving SNAP benefits in 2017.

I'll say that again: **40 million Americans**. To get your head around such a large number, the population of Canada as of July 2021 was 38 million. In Australia, a country of 24 million people with a 10% average spend on food, over five million people experience food insecurity yearly, and one-quarter are children.

To simplify the 828 million people globally or the 40 million Americans, think of it as **one in ten people undernourished**. As I said, next time you are on a bus during rush hour, look around. At least 6 of the people on the bus represent the global undernourished. Then add another 2 to cover those malnourished from too much of the wrong food. One in ten people and one in six kids should be unacceptable statistics.

And hunger is so easy to fix. Check his quote from economist Kate Raworth's excellent book *Doughnut Economics.*

Food consumption is deeply skewed too. Around 13% of people worldwide are malnourished. How much food would it take to meet their caloric needs? Just 3% of the global food supply. To put that in context, 30%–50% of the world's food gets lost post-harvest, wasted in global supply chains, or scraped off dinner plates and into kitchen bins. Hunger could, in effect, be ended with just 10% of the food that never gets eaten.

Kate Raworth, Economist

It is staggering if what Raworth says is true that 3% of the global food supply would feed the undernourished and that the percentage is so low because **just half of the food produced gets eaten**. Why would we grow so much if it doesn't get eaten? It seems bizarre.

Reliable evidence for these numbers is hard to establish because of the complexity of the global food system. However, there are many opportunities for spoilage and loss from harvest to transport, storage, processing, and retail. Once the consumer obtains the food, there are more losses in storage, cooking and what gets left on the plate by the picky child unable to eat his greens.

Broadly, losses happen as agricultural produce is transported, processed, and prepared in the global food supply chain. Food losses for the 2 billion or so people in 500 million households living in rural areas of developing nations who survive as "smallholder" farmers, growing their food on less than 2 ha, happen as the crops and livestock are produced and stored. On the farm or in the supply chain, the opportunity for waste exists.

Everyone must eat. We all need food to be available, accessible, usable and reliable to be food secure. As the fossil fuel energy pulse begins to fade, 6 billion rely on the global food supply chain to provide this food security. It is a precarious situation, yet we allow food waste and hunger to be in the same sentence. Sure, have a go at the politicians and the corporations we let behave in ways that got us into this unholy mess. But remember that the social and economic systems in which these villains thrive are of our making.

Suppose the food waste and hunger connection is out by a factor of 3, and the real number to feed the hungry is 10% and not 3%. Given this conservative estimate, cutting the food waste numbers by a third would provide enough food. This should be a colossal embarrassment to everyone who pays less than 20% of their income for food.

If you go to bed on a full stomach, consider yourself one of the lucky ones. Many people are not so fortunate. Be mindful of your food waste. And if you are one of the lucky ones, be grateful for the meals you enjoy and never take your food security for granted. Be embarrassed, then be angry and demand better.

Thermodynamics

Unless it is tended, a swimming pool will eventually become an ugly, green puddle. Natural processes result in decay because they must comply with the Second Law of Thermodynamics, which is the tendency of natural processes to lead towards spatial homogeneity of matter and energy, especially of temperature. Physicists measure this with entropy. In an isolated thermodynamic system left to spontaneous evolution, **entropy** cannot decrease with time, and **isolated thermodynamic systems always arrive at a state of thermodynamic equilibrium, where the entropy is highest**. In short, nature decays. Even great mountain ranges are eroded to flat plains, weakening the power of water to erode. But here is the fantastic part.

Biology is a system to defy entropy, at least for a time. Organisms are thermodynamic systems that stave off entropy using energy to put off the inevitable thermodynamic equilibrium for as long as it takes to grow and reproduce.

Organisms maintain complex internal and external organisations using energy to do it. A plant does this by capturing energy from the sun. A rabbit does it by eating the plant. Humans do this for our bodies with energy from food, but we have also managed to capture exogenous energy, first with fire and then fossils. Extra energy maintains our environments for the best comfort and convenience. Humans are exceptional at defying entropy until we age and falter.

The first law of thermodynamics states that **when energy passes into or out of a system (as work, heat, or matter), the system's internal energy changes in accordance with the law of conservation of energy**.

The law of energy conservation states that an isolated system's total energy remains constant; it is said to be conserved over time. Energy can neither be created nor destroyed, only transformed or transferred from one form to another. If defying entropy uses energy, then that energy must go somewhere. It dissipates into the environment.

Organisms are waste machines. When they maintain themselves, they dissipate heat, and when they decay, they dissipate heat. Because the earth is finite, it cannot be a bottomless pit for waste materials or energy. The organism does not perceive the waste. In real time, it just happens, apparently without consequence. Nor does the organism perceive the finite.

An impossible nirvana

Until 12,000 years ago, nature was a vast human playground, an endless landscape of opportunity. There was no way a tribe of our hunter-gatherer ancestors could imagine there was a limit so long as they kept moving. A handful of the lucky ones living on the shores of lakes or river banks or next to the ocean did not even have to move very far.

What they knew was daily, seasonal and yearly change. Nature was cycling around them, coming and going with the tides and the weather. Water flows over the waterfall and always will. The falls cascade only after storms or during the rainy season, and it can be down to a trickle in the dry season. Severe winters might cause the water to freeze. There is no constancy then, but water flows over the waterfall.

When humans invented agriculture, little changed initially. A few sedentary cultures came and went, but most humans still had to move around on what must have felt like a vast, empty planet. Resources were scarce, but they replenished themselves, and the water went over the waterfall.

Then, when agriculture became the dominant source of food—the energy source to power human bodies so they could defy entropy—the fields and grasslands became net energy sources. The sun's energy and some human and animal labour went in, and the food came out. At a planet scale, this was similar to what happened before agriculture when the vegetation fixed the sun's energy and the animal biodiversity processed and cycled it along with nutrients. In principle, agriculture that mimicked nature as a net energy source with internal cycling of nutrients could persist in ecological time. It wasn't all that different to the ecology that had been going on through evolutionary time since plants first colonised the land 470 million years ago.

The second law still applies—isolated thermodynamic systems always arrive at a state of thermodynamic equilibrium, where the entropy is highest—but nature had been defying this law all along by cycling energy and nutrients and absorbing the energy waste. It is the miracle of life. In ecological times, nature appeared consistent, stable even. But the reality was that the waste released from all the energy used was filling up the earth's capacity to absorb the excess energy. Early agriculture changed the balance. Some additional waste energy dissipated into the atmosphere when the vegetation was cleared for the fields. Most fields are less productive and less ecologically efficient than natural vegetation. They are also designed to export energy and nutrients to the human population, who increasingly consume them away from farms.

Then came the Haber-Bosch process and the mechanisation of agriculture to deliver an exponential increase in the use of nitrogen fertiliser. This new, intensive agricuture was a huge energy sink. More energy went in than came out as food, even though the amount coming out was huge and was converted into a population spike. However, thermodynamics still applies to the energy added, and it leads to thermodynamic equilibrium.

Where did the waste energy go? Into the sink that is the planet. It had nowhere else to go. And almost all humans are blind to it and ignore thermodynamics, especially the second law.

Key points

- Industrialized food production systems, like China's intensive pork industry, have been able to increase output massively but rely heavily on inputs and technology that create potential vulnerabilities.

- Key resources like phosphate fertilizers have highly concentrated sources that pose supply risks, even if total reserves seem sufficient. Geopolitics around critical resources is an underappreciated issue.

- The staggering levels of global hunger and food insecurity co-existing with equally appalling levels of food waste across the supply chain is a moral and practical failure that urgently needs addressing.

- Human activity fundamentally defies entropy by harnessing energy, but our systems are now overwhelming the planet's capacity to absorb the waste, violating thermodynamic limits.

In the next chapter

As speculation drives agricultural land prices to dizzying heights disconnected from their inherent productive value, the very foundation of our future food security is being undermined.

Chapter Three

Land prices

As speculation drives agricultural land prices higher, disconnected from their inherent productive value, the very foundation of our future food security is being undermined—it's a worry.

Real estate

In Australia, suburbia runs on house prices rather than land prices. The weekend talk around the Webber BBQs, resplendent with their peeled prawns, often ends up in the challenge of paying the mortgage. You know, the massive sacrifice taken on by each family to secure the picket fence because, in time, the house price goes up to make it all worth it. And so it has been in the 27 years I have lived in the great southern land. I am one of the lucky ones. Our house in the Blue Mountains west of Sydney has nearly trebled in value. Whilst we were paying for it, it was an epic investment with risk and hefty monthly repayments, but for now, we have a tangible asset.

I am old enough to remember the highest-ever UK interest rate of 17% in November 1979. I was a naive freshman at the time with no thought of house or land prices. My wife remembers, though. She was the proud owner of a mortgage in Australia during the 1980s when interest rates were in the teens. Back then, the borrowings were modest, and the repayments were manageable. Today, you must add a zero to the borrowed amount, which changes the game. At the end of 2021, the interest rates in Australia rose far enough for the average mortgage repayment to grow by $1,000 a month before 2022 clicked over. To put that into context, the Australian Bureau of Statistics in August 2021 had the average weekly earnings for all employees in Australia at $1,305 (or $67,860 per year). $1,000 extra a month is a staggering reality that scares the pants off the average John Doe. Buying the house with the picket fence costs another $15,000 a year before tax. There will be mortgage defaults. There has to be.

Average arable farm size in Australia is 4,750 ha with an average value of $9,000 per hectare makes a rural property worth over $40 million.

Should we choose to sell our suburban house, the market will set the price for the dwelling. A potential buyer wants a place to live, the house. The price reflects what people in the market are prepared to pay for that utility. Our house sits on a little over a third of an acre—1,528 m^2—only the buyer is not concerned about the land beyond how it frames the house and if it is in the right location.

A farm used to be similar. The price of a farm reflected its capacity to grow food and make a profit. Buyers of farms were interested in the business. To have a successful farm business, you need land, typically a few thousand hectares in the dry southern land. So, rural real estate is about buying land, not the business running on it. Today, that disconnect is acute.

> *Land prices continue to soar, but this phenomenon is not an indication of improving profitability or resilience in the agricultural sector. It reflects a genuine decoupling of the agricultural real estate business from the agricultural production business. And it will precipitate a necessary structural adjustment that has the potential to be very destabilising to both businesses.*
>
> Grain and cattle farmer, northern NSW, Australia

The value of the land has become far more significant than its potential to grow food. Why is this?

The value of land was about what you could do with it. For example, land suitable for crops, let's say growing wheat that yields around 2 t/ha and sells for $300/t, is worth more than rangeland that can't be used for cropping but can carry 8 DSE/ha and might have a gross margin of around $250/ha running cattle. Wheat makes more money than cattle, so the land that can grow wheat is more valuable.

Plus, to make the cattle gross margin, there is a minimum number of cattle in the herd and a minimum land area needed to raise them, at least 150 ha in the wetter parts of eastern Australia, more as you go inland. Moreover, wheat might make a lot of money in the good years. Combining an above-average yield with a spike in the price per ton and buying the arable farm makes perfect sense.

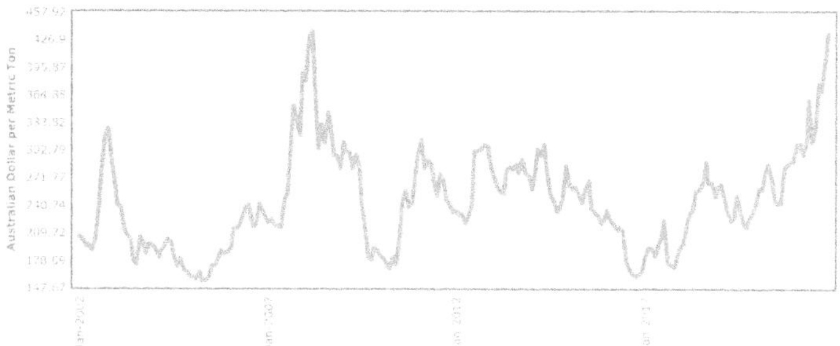

Wheat prices have fluctuated over the last 20 years with steep falls and sudden rises that reflect global demand and supply volatility. A good crop in 2003 or 2008 or again in 2022 is a massive earner for grain growers.

While the grain growers dream, the animal producers trudge along without the bumper years. They benefit from more reliable returns but don't get the windfalls from the wet years and still suffer from droughts. Scale is what makes money from livestock. Historically, these production factors mattered when agricultural land was valued. Then, something more substantial came along.

The median price per hectare of Australian farmland increased by 13% in 2020, the seventh consecutive year of growth, bringing the 20-year compound annual growth rate (CAGR) to 7.6%. Although notoriously bumpy, the average stock market return is about the same. The S&P 500 returns with dividends included had a CAGR of 5.7% over the same period. Land is a good investment again.

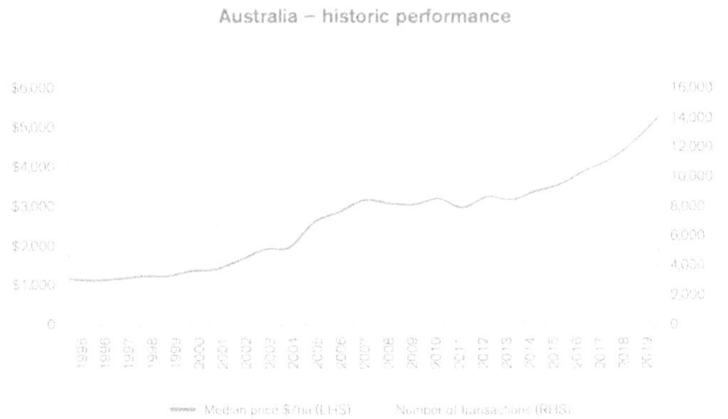

Image source: Australian Farmland Values 2020 New South Wales, Rural Bank

The 2020 13% year-on-year price growth is a heady hike, with analysts pointing to a combination of reasons: high commodity prices, historically low-interest rates, good seasonal conditions, and intense competition for prime farmland as farmers seek to expand their operations. It's a spike. The steady capital growth—land price from $1,000 to $14,000 per hectare—over decades is different, and it's not because the land makes more money than it did. Indeed, profit margins are tighter than ever because input costs have risen sharply. Agricultural land price gains are about scarcity and capital markets flush with cash looking for a return.

Agricultural land is in demand because investors and governments worldwide have realised that **good farming land is about to become a critical resource.** Access to land is a vital part of any forward-thinking food security strategy. This is partly about the absolute supply of food for urban populations and partly about the ability to buffer supplies of foodstuffs to avoid market volatility and panic buying. It is one thing to have a shortage of toilet rolls and quite another to have price hikes on staples like bread, eggs and dairy. Toilet paper shortages might lead to fisticuffs in the supermarket aisles, but food shortages can incite riots.

Land prices set for the scarcity of agricultural land rather than intrinsic production value is a global phenomenon because countries with high food demand and deep pockets pay close attention to food security; they must.

In 2019, there were 66 million sheep in Australia, roughly 2.6 sheep for every person.

Housing affordability in Australia is already a critical social challenge. The urban dream of a house, yard, and white picket fence is now a myth. Even for DINKY couples, average wages cannot service seven-figure loans no matter how frugal the household. Interest rate pain rubs in this reality. Decoupling land's intrinsic and utility value means the same happens in land prices in rural areas. Beyond the sovereignty issue, rising land prices are impacting succession planning for family farms and putting up a massive barrier to new entrants to farming. At least two things follow.

The first is the trend for the consolidation of farms. Larger farms can raise the capital to get bigger by buying up their smaller neighbours. Banks and lenders are good with this because they are investing in the land asset that, on its own, generates a competitive return. The farmer gains from economies of scale. This is the inevitable trend to intensification that is great for yield but not for collective over-reliance on inputs. Intensive agriculture is necessary for feeding everyone well, but it tempts landowners and their land managers to push the land hard. Intensive farming relies on inputs and moves away from natural capital.

The second is the need for more rural residents. One of the few ways to wean ourselves off energy subsidies while still growing enough food is to build farms with closed-loop nutrient exchange through systems like regenerative agriculture, restoration, and organic production. All fancy words to describe more people to supply the energy to run the production system—people and horses, livestock and cover crops to replace at least some of the fossil energy inputs.

These people currently live in the cities because, despite the high cost of living, it is where the supply of services from education to health and entertainment is most reliable. It turns out that rural land is a social issue. All this change in land price is in motion. The market, tweaked by regulation and government intervention, will continue to operate through supply and demand to set land prices. It is what it is: a decoupling of values. That is what markets do. They develop and evolve as the people who make trades in them decide on what is valuable today and what could be more valuable tomorrow. It's a game.

The land has always been valuable. People desire it and have gone to war to acquire it, so it will be valued in markets. The problem is that we should value the land's ability to grow food and sustain ecosystem services much more than a dollar return. Making land more valuable than the food it produces is not the only threat to our economic system, but it's a huge one.

By the way...

In Australia, one DSE, the shorthand for Dry Sheep Equivalent, is the feed energy required to maintain a 45-kilogram live weight Merino whether with zero weight change, no additional wool growth included in maintenance, and walking 7 km/day. 1 DSE has an energy requirement of approximately 8.7 MJ ME/day, and all other livestock are benchmarked against this value. Typically, a yearling steer is 8 DSE.

Key points

- Land prices, both for housing and agriculture, have been rapidly rising in recent years, driven more by speculative investment demand rather than the land's intrinsic productive value.

- For agricultural land specifically, factors like low-interest rates, funds seeking returns, countries pursuing food security, and competition from expanding farmers are decoupling land prices from farm profitability.

- This trend encourages consolidation into larger industrialized farms and intensification reliant on inputs while pricing out smaller family farms and new agricultural entrants.

- We need to reframe how we value land, prioritizing its ability to sustainably produce food and ecosystem services over treating it primarily as an investment asset class.

In the next chapter

8 billion people, increasing at 8,000 an hour, have needs for housing, schools, hospitals, and other components of a modern economy. Much of this infrastructure required by 2030 remains unbuilt. There are monumental challenges in meeting these needs while considering ecological costs and societal well-being.

Chapter Four

$4 trillion of infrastructure

More than $4 trillion of new infrastructure is needed worldwide each year. We are so good at using resources that we should not be surprised by any numbers. We are more making machines.

A huge number

$4,000,000,000,000 is a huge sum. It represents $500 for every man, woman and child on the planet. So why baffle you with an unfathomable number?

Annual $4 trillion in global spending is required to meet the demand for housing, schools, hospitals, and all the paraphernalia of a modern economy. And the rate of development will need to keep going each year. As of 2017, over half of the urban area needed by 2030 has yet to be built.

Around 60% of the area expected to be urban by 2030 has yet to be built so the technologies used to create that infrastructure will have far-reaching social and ecological implications.

Kate Raworth, Doughnut Economics

As the built area grows, so do the paved roads, bridges, pipes, cables, and a plethora of infrastructure that modern society requires. Most of the development will happen in Africa and parts of Asia that will see the highest population growth. All this is too vast to comprehend. It also assumes sufficient economic prosperity or debt to pay for it, along with energy and physical resources to realise it. But even if the spending declines or needs are unmet, by 2050, the world will have more built area than it does today. I have been around long enough to know exactly what this means.

Africa is not a country

I remember a visit I made to Uganda in the late 1980s. Maribou storks roosted on the tall buildings in the capital, Kampala, and the potholes in the city's roads swallowed vehicles whole, but the people smiled wide and warmly welcomed the white scientist. The rickety Kenya Airways 707 that brought me into Entebbe airport from Nairobi parked next to a gleaming British Airways 747, an alien creature that dwarfed everything, including the terminal, a tiny concrete building. Today, the terminal buildings sprawl across the landscape. The airport is still small by international standards but has grown over the years as part of global infrastructure spending.

Entebbe airport has come a long way from the one depicted in the action thriller movie, Raid on Entebbe produced in 1977.

Imagine telling that anecdote to the villagers of the region 150 years ago, already shocked by their encounter with John Hanning Speke and Richard Francis Burton in 1856, who were on their East Africa expedition to find the Great Lakes. Metal birds, concrete and buildings made of stone were as alien as the pale-skinned men with weird gourds on their heads.

Colonialism, religion, and racist ideology were not the only things white men brought to Africa; they also carried the knowledge of infrastructure with them. It is Africa where much of the 60% urban expansion Raworth quotes will be needed. And it happens so fast. Blink, and you will miss it is the message here. Infrastructure is going up everywhere at dramatic rates. Even in developed countries, some changes might not be obvious but go away for a while, and even established cities look a bit different when you return. Sydney now has a network of road tunnels, including the second-longest in the world, and most of them were just an idea when I arrived in the country from my African adventure in 1996.

Botswana is a small country with just over 2.36 million people in 2020. It has few of the population problems that trouble most African nations. But when I went to work there in 1989, just 30 years ago, there were 1.24 million people.

The people of Botswana are big on education. The teacher is as revered as village elders, and the government has invested in education, including new schools. Around 27% of the population are school-age children so, in the 30 years since my time there, Botswana has needed an additional 300,000 school places. Assuming each school has 1,000 students, this means 300 new schools to keep all the children in the classroom.

This is the blink.

Nigeria's population was 93 million in 1989 and 206 million in 2020. Assuming similar proportions of school-age children, the Nigerian government had to build infrastructure for an additional 79 million students or 79,000 schools. This was a blink but also a monumental challenge.

Accra is the capital and largest city of Ghana, with a population of 5,455,692 inhabitants in the 2021 census.

A key point that Kate Raworth made in *Doughnut Economics* is that how this infrastructure is built has a huge impact on the environmental costs and the well-being of the people who inherit the buildings and roads. Schools are needed, but what type, where and with what materials? Answers considering the context, materials and risks will make a difference to future resilience and prosperity. For example, the International Institute for Sustainable Development (IISD) found that using natural infrastructure to protect against climate breakdown could save up to $248bn (£180bn) a year globally, costing only about half as much as equivalent built infrastructure and delivering the same protection. 'Natural infrastructure' refers to natural landscape features and/or nature-based solutions that promote, use, restore or emulate natural ecological processes.

All very nice, but the reality is that most of the new infrastructure will be similar to the old, made of concrete, steel and tarmac, embellished with plastic and a dazzling array of petroleum products, at least while there is still affordable oil available and financial markets with investable cash.

Don't be surprised

All the time, we are told that extinction is happening, that habitat is being destroyed, and that the environment is under threat at every turn. Because it's so common and we feel it and hear it every day, this rhetoric makes us numb. It's all so sad and painful and, well, too much. The interesting thing is that we shouldn't be surprised by all the bad news. Nor be confused or doubt the severity of the impact of humanity on the planet. Just think for a moment what it takes to maintain a single human life, your life, for example.

Every day, you need to take in some water, be it in the 3-litre fresh allocation, several coffees, a soft drink or even a few beers. Liquid has to be taken in for the body to function. It is also essential that you excrete some of the fluids going in because the kidneys are using that water to take away some of the products of metabolism, the waste that your body no longer requires.

Everybody has to eat because, without the energy and nutrients from food, the body can't function. You need fuel to power the cells in your body, so consumption of food is a requirement. Again, not all of that food is absorbed. Some of it is excreted, so there is a requirement for that function, too.

Then, of course, you need to find something to do all day to acquire those resources that maintain your body. In ancient times, people would go out and forage for themselves. In modern times, we use the marketplace, occasionally growing a bit of our own food on the side. About 2 billion or more people around the world grow all their food because their low income means they have little choice. Around 6 billion of us trade some or all of the time that was spent either gathering or growing food and water for money—this money we use to purchase those essentials.

In nearly all of us, there is an innate desire to reproduce. So we look for a mate and form a family unit. It is so pervasive this desire to make more that all around the world, humanity has created rituals around formal unions to support a family. Once the babies appear, more food, water and shelter are needed, somewhere safe where the children can grow and be nurtured. And even if you flout your genes and choose to go childless, a roof over your head makes life much more comfortable. Every one of us on the planet does one variant or another of these things as part of our innate biology.

As large mammals, we have a significant resource requirement for maintenance and to make more. We take what is needed from the vegetation and the soils, and each of us has an ecological footprint of one size or another. Modern technology has extended this capability for resource gathering beyond people's wildest dreams. The fact that we now have electricity, houses, vehicles and an extraordinary infrastructure to support a modern lifestyle provides the essentials only with considerably more effect and a larger footprint.

It took 450 words to describe these life essentials. At an average reading rate of 140 words a minute, it has taken you around about 3 minutes to read them. In that time, roughly **450 people have been added to the global population**.

This happens every minute of every day — 130 people per minute are added to the global population at our current rate of growth, that is, births minus deaths in each minute. Each of those 130 people will grow up and require a similar kind of ecological footprint to the one described. They will need to eat and drink and have shelter and trade their time for money for a life that might extend to 70 or more years.

Over one hundred people a minute are added to a population of 8 billion souls. When you try to understand these numbers, it should be no surprise that humanity is appropriating the planet, taking up resources at an extraordinary rate.

Kecap Manis

Suppose you make a delicious fried rice dish with an egg on top. It is already enticing but made even better with some sweet soy sauce and kecap manis, as the Indonesians call it, to complement the rice. The condiment comes in a small bottle, and only a little is needed to finish the dish. If there are two of you, then one small bottle sits in the fridge and lasts for months. Not every day do you have special fried rice, and two people don't need much sauce.

Suppose the kids come around for supper, bring their kids and a couple of mates. Now, there are ten people around the table, and they all decide that they would like some kecap manis. The bottle will be consumed in one sitting. This is the reality of numbers.

It's not necessarily the size of the footprint of each person, although this matters; it's the collective footprint of so many people.

- 130 a minute
- 8,000 an hour
- 192,000 a day
- 5,840,000 a month
- 70,080,000 a year

Humans are insatiable. We cannot help ourselves in betterment. Getting and being more is hardwired into our psyche because back in the day, if we slacked off, we perished. We can survive with less and even find true happiness doing it, but the basic needs must still be met. And we currently do this with infrastructure. Indeed, we could not meet the basic needs of 8 billion people without it. Building it will cost money and use resources.

Rather than limiting reproduction or pretending to be an ostrich, it is better to wake up to reality. We have to feed everyone now and keep it going.

Neediness

Food, water, and shelter are essential to human life. Fail to obtain one or more of these fundamentals, and the complex chemical engine that is our body succumbs to entropy, and we die. In modern times, we also need each other because no one person has all the skills to gather enough food, water, and shelter to avoid the perils nature coughs up. We all need help.

Neediness that becomes a handicap sending a person to a therapist is an expression of this fundamental reliance on each other. We function well together because our solution to defy entropy is creating innovation and technology to gather and exploit nature's resources. This is best done in groups or, if you prefer, teams.

We could all transcend to the virtual mountaintop and meditate away our egoic selves to leave our needs behind. Then death would stalk us all because human life is only maintained with food, water, and shelter. Even the monk on the mountaintop has to eat and drink. He also can't Wim Hoff his breath through the inevitable mountaintop blizzards. He'll need some shelter. Meeting these fundamental needs for 8 billion people is a 22 trillion calories-a-day task.

What we often forget is that fossil fuels gave a vast energy pulse. The extra people we created from them are between 4 and 6 billion more than unmodified photosynthesis can support. Feeding everyone alive will require an energy subsidy to channel extra photosynthesis into food.

Almost all significant human activities are linear and irreversible—mining minerals, oil, gas, coal, soil, groundwater, ecosystems, biodiversity, etc. We are not good at cycles.

Humans also dump toxic chemicals everywhere in the environment (atmosphere, water, and land) that change the cycling conditions. And keeping the thermodynamic theme, one of the primary pollutants—greenhouse gases—also makes it harder for the excess heat to escape into space. Everyone knows all this after just a moment of consideration.

The stuff that we buy, including food, comes from somewhere else, is consumed by us, and then thrown away or flushed down the toilet. It leaves us to go somewhere else. Maybe we think a landfill is just a hole in the ground that the garbage trucks fill up or that the sewer takes waste and miraculously cleanses it at the water treatment plant. It doesn't.

Energy use is biologically inevitable. Only we have added exogenous energy from fossils to turbocharge resource use. The energy needed to make all the goods and provide all the services demanded by humanity generates waste as physics dictates. Just ask the oceans what they think about plastic.

> *no scientist can claim that the human economy is sustainable*
>
> Bolson, N., Yutkin, M., Rees, W., & Patzek, T. (2022). Resilience rankings and trajectories of world's countries. Ecological Economics, 195, 107383.

Why we shouldn't expect sustainability

Humans use a staggering amount of energy every day. We need endogenous energy to power our bodies—around 2,500 kilocalories from food. Then, there is the exogenous energy used outside our bodies to make our lives easier. Along with the 22 trillion kilocalories for food, in 2019, primary energy consumption was 173,340 terawatt hours. Roughly 80% of that energy came from oil, coal and gas.

Even the adjective exogenous gives it away. Humans are the only organisms that can do this. It started with the control of fire, was helped with the invention of the wheel and the domestication of animals—it's not called horse power for nothing—and went ballistic with fossil fuels and the technologies they power. And to put that exogenous energy consumption into context, back in 1800, the consumption was 5,653 terawatt hours.

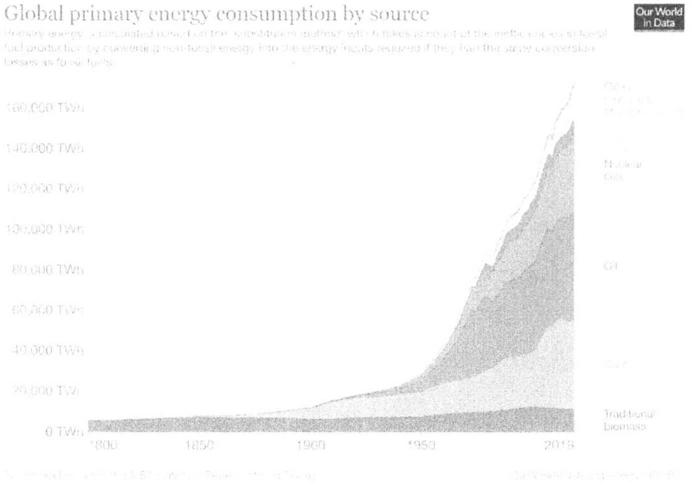

Exponential increase in the human consumption of exogenous energy since the industrial revolution

Exogenous energy is the source of our consumption. It grows our food, makes all the goods we consume, and creates the technology and infrastructure that make modern lifestyles possible. What we get exogenous energy to do is miraculous, and it happens in an instant.

In 1819, the city of Memphis, Tennessee, was founded, Alabama was admitted as the 22nd U.S. state, and the African Slave Trade Patrol was founded to stop the slave trade on the coast of West Africa. Just two hundred years later, humans went to the moon, sent probes across the solar system, and erected a telescope in space that could see the galaxy GN-z11, which is located about 13.4 billion light-years from Earth.

Two hundred years is less than 0.07% of our time on earth as a species. If the 12,000 years since we invented agriculture is condensed into an hour, the time from 1819 to today would be one second!

Using exogenous energy is an unprecedented leap in human evolution and cultural development. A kid has been let loose in a candy store with no adults. The kid goes on a sugar high, gets a stomach ache, and his parents send him to the psychologists to prescribe some ADHD medications. What should we expect will happen? Growth in energy use is the kid in the candy store. As Nate Hagens puts it in his paper on the superorganism

> *If the global economy continues to grow at about 3.0% per year, we will consume as much energy and materials in the next ~30 years as we did cumulatively in the past 10,000*
>
> Nate Hagens

Humans are instantly addicted to this exogenous energy bounty that allows billions of us to live like kings for minimal personal effort. History tells us that moral outrage led to the abolition of the slave trade, but it might have taken longer without the energy and technology to do all the heavy lifting.

Energy use correlates directly with GDP, almost at unity. Energy use matches resource use that fuels economic growth. This correlation is not in dispute. It is logical because of the need to convert resources into goods and services bought and sold, none of which is possible without energy or, in earlier times, slaves.

As a highly successful species, we use resources, both natural and man-made, all the time. And we have energy slaves to help us do it. Global oil consumption is roughly 100 million barrels a day. This number is a bit rubbery based on what definitions are used and can be nuanced until even the oil men get bored. Still, the known reserves of oil can meet this demand for roughly 47 years—roughly one and a half future generations.

Then, there is the resource use—minerals, land, vegetation, and soil—made possible by readily available energy to power machinery to make food and consumer goods. This depletes resource stocks to the point of compromise. We know this. A high school kid can figure it out. But we choose to ignore the obvious.

Neediness is the human condition. We are needy beings because, at a primal level, we have to find fuel to maintain our chemical engines, or we die. But there is more than basic metabolism. Humans fine-tune our miraculous talent for resource use to make our lives more comfortable because physical comfort is easier on the body's chemical engine. No matter that comfort is corrupted by the drive to achieve it, we don't know when to stop adding more comfort. And this neediness is insatiable. It will not go away.

Then we found fossils as a miraculous external energy source and set about converting that energy into more people. That pulse of extra people, several billion of them, who are here because we tapped an external energy source, makes the future of humanity impossible without 1) maintaining an eternal energy source and 2) curbing our neediness.

Our best hope is to find another clean energy source to keep the human engines running until neediness is mastered and our population declines back to the limit imposed by planetary boundaries. Wish everyone luck. A society that runs on exogenous energy with technology capable of converting that energy into an array of creature comforts needs a big planet... Or six. Unable to escape our basic biology to make more, we use that energy to grow people and wealth. It is an innate response. Now that we know how to use exogenous energy, we should expect resource use and depletion because there is nothing in our instinct to stop us. The only thing that will is if energy or resources run out. At the rate we are going, that will be sooner than we would like.

I will leave the late, great E. O. Wilson to sum up the thoughts in this chapter succinctly.

> *The real problem of humanity is the following: we have paleolithic emotions; medieval institutions; and god-like technology.*
>
> Professor E.O. Wilson, American biologist, naturalist, and writer

Key points

- Over $4 trillion annually in infrastructure spending is required globally, primarily for housing, schools, hospitals and urban development in rapidly growing regions like Africa and Asia. This represents an unprecedented construction boom.

- Much of this new infrastructure is being built extremely rapidly, with entire new cities and infrastructure networks being built in just decades or years—a "blink of an eye" pace that makes the environmental impacts difficult to manage sustainably.

- If constructed using traditional materials like concrete, steel and plastic, the resource demands and ecological disruption from this $4 trillion infrastructure wave will be catastrophic.
- The severity of these environmental impacts should not surprise us, as even sustaining one human life requires significant resource consumption multiplied across a global population of 8 billion.

In the next chapter

Consuming the finite planet's resources with abandon is as unsustainable as eating through a limited stash of breakfast money without replenishing it. Let's see what that looks like.

Chapter Five

Breakfast

Consuming the finite planet's resources with abandon is like eating through a limited stash of breakfast money without replenishing it.

Breakfast at Macca's

Back in the day, there was no intermittent fasting for me. I needed to eat breakfast. I was a Hobbit when it came to the day's first meal. So imagine my delight as I passed the local McDonald's advertising a one-dollar special, McMuffin and a coffee. It's a miracle that food can be so cheap.

Back at home, I have a stash of dollars, a thousand of them, making my breakfast needs easy to satisfy. I pick up a dollar bill each morning, put it in my pocket, and hand it to the polite teenager behind the Macca's counter. She passes back my breakfast, and off I go to work. The wheat and caffeine kick in, and I feel great... until coffee time.

I have a thousand one-dollar bills sitting in my apartment. My stash will last a long time. Given the way fast food outlets operate, the special will no doubt continue, so there is no need to worry. Breakfast is a given. A thousand days is beyond anything I'm prepared to imagine. Several years of breakfast are secure.

After a couple of years of tasty McMuffins, I realise that the pile of dollars that was so tall it looked as though it would last forever, well, it is considerably smaller than it used to be. I did not replenish the stash, so I'm looking at my last few months of breakfast. I dither and worry but do nothing to add funds. And sure enough, one cold winter morning, I'm down to my last dollar. As I handed it over at McDonald's, where the special is now a bacon burger, I had no idea where tomorrow's breakfast would come from.

Despite the dubious nutritional credentials of breakfast from a fast food outlet, the story here is simple. **Vast resources at the beginning will run out unless they are renewed**.

Money is not automatically renewable unless you are a central bank. Normal folk must replenish any stashes of cash from earnings or interest on investments. Perhaps money is too fickle a resource. Instead, let's think about apples.

Suppose I eat apples from an apple tree in my garden. If I'm looking after the tree, pruning, applying fertiliser, and keeping the birds away, then in theory, that tree will produce apples until it reaches the end of its expected life. And if I replace that tree before it gets too old and tired, I can, in principle, consume apples in season indefinitely.

When I eat the apple, some energy is burned in my cells, and the nutrients might make their way into my tissues. The bulk of the apple is not fully digested and ends up in the toilet, where it is flushed into the sewer. I am a diligent horticulturalist, and I look after my trees with mulch and regular horse manure application. Any nutrients I apply to the soil replenish those nutrients that the plant removes to make the apples I consume and send out to the ocean. The addition of fertiliser, water and the occasional pesticide application must not affect soil properties. If the soil compacts or becomes saline or loses organic matter, then adding nutrients may not be enough to maintain fruit production. These changes are likely subtle and take many decades to affect the tree.

Indeed, the apple trees in my garden may persist for generations but not indefinitely. At some point, conditions become different enough for another crop to be more profitable than apples. Even the apple tree that grows for decades and fruits each year requires the addition of nutrients as fertiliser and not treating the soil like a factory. My apple orchard receives these nutrient inputs via fertilisers that had to come from somewhere. And so it is with open systems. The inputs have to come from somewhere; outputs cannot persist indefinitely without them.

Estimates are that 7,500 varieties of apple are grown across the world

Coal

In the 1880s, when coal was first used to generate electricity for homes and factories, it was a revelation. Limitless cheap electricity and engines that could pull carriages of goods and people for mile after mile was an economic miracle. Coal provided power the Victorians believed could take them around the world, and it did. They didn't contemplate that such a remarkable and profitable resource would run out. There was so much of it in the ground, and if you went to other parts of the world, you would find it everywhere. Coal would be the source of power forever.

Even when oil created vast new opportunities as a liquid fuel, the wonder of coal kept its shine thanks to electricity. The Edison Electric Light Station in London was the first coal-fired power station built in 1882. By 2019, there were an estimated 2,425 coal-fired power plants globally. More dramatic than this 140-year growth in the use of coal is that 50% of these power stations came online since the year 2000. Here is what happened to coal production after the men in top hats went in horse-drawn carriages to the first ballrooms lit by electric lights...

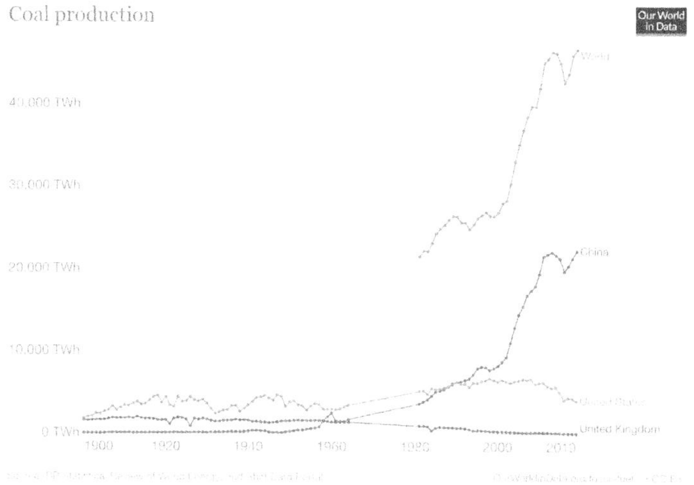

Coal production worldwide has increase dramatically this century

Today, of course, fossil fuels are less shiny, given their impact on climate. Those 2,425 power stations emit around 15 billion tCO2e. A rush is on to alternative fuel sources without greenhouse gas emissions for some stability in the planet's atmosphere. Even China, which gobbled up most of the growth in coal use since 2000, is cutting back with a commitment to stop funding coal power overseas. There is talk that we have reached or even passed 'peak coal' as coal's share in the global energy mix was highest in 2008 when coal accounted for 30% of global energy production.

In the beginning, coal use was modest—just a handful of steam engines, power stations, and some fireplaces to generate smog. Globalisation was in the future, and the population in need of power was a little over 1 billion. It is hard to consider shortage when demand began modestly, and supply was easy enough to dig up out of the ground. It turns out that if the Victorians had used coal at the rates we do today, all the reserves would be gone by now.

Of course, coal is a nonrenewable resource. Mining depletes the stock, and the time it takes to make coal is way beyond anything human lifetimes can understand. Burning coal releases greenhouse gases that are an externality, a cost of use not paid by the user. Coal is not renewable; we mine it.

Coal reserves are depleted, but not to the point that they have run out. If the reserves are large relative to extraction—there is a similar issue with groundwater—but extraction is called mining for a reason. The Victorians built an industrial economy with no notion it would ever end. Why would they? The resource was as big as a pile of one thousand dollar bills for breakfast. No matter, we have alternative sources, such as green ones.

There are roughly 2,435 coal-fired power plants worldwide, almost half of them in mainland China.

Coal-fired Teslas

If you drive a Tesla Model S for 100 miles, it will use roughly 34kWh of electricity. In petrol terms, that is about 100 MPG and sounds like a big win for the driver. For example, let's say electricity costs an average of $0.12/kWh, so the yearly cost to drive a Tesla Model S 15,000 miles is $612. Not bad and much cheaper than gasoline. Driving those 15,000 miles of silent motoring pleasure needs roughly 5,000 kWh.

But what about the carbon emissions? A new petrol car would be expected to release around **4.4 tCO2e** to cover the same yearly distance. Surely, an electric car would have fewer emissions.

Electric engines are more energy efficient for the atmosphere than internal combustion engines. Using electricity to drive the car means no exhausts and no nasty fumes. Aside from the embodied energy in the materials that make up the vehicle, the carbon footprint should be minuscule. Perhaps. The thing is, it depends on where the electricity comes from.

Over 56% of the electricity in the U.S. comes from coal-fired power stations. More than 84% of the coal consumed each year in the U.S. is used to generate electricity; in Australia, it is 75%. It will take a while before most energy generation is from renewables with a low emissions footprint. In the meantime, at least half the electricity will be from coal.

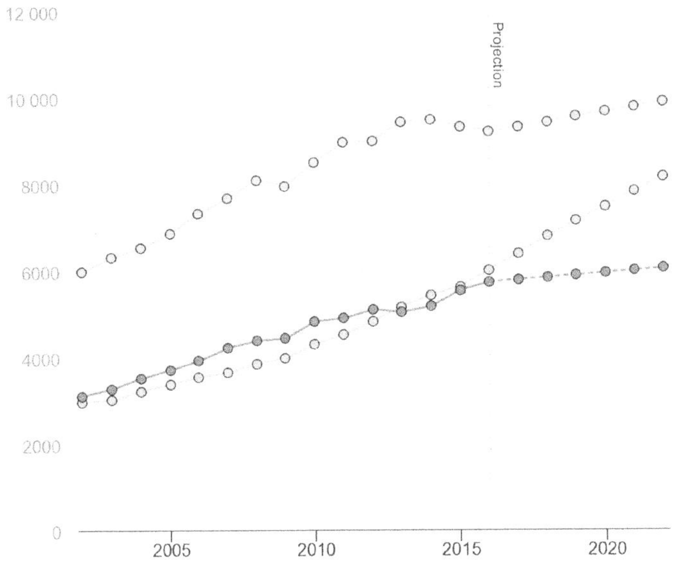

Global electricity production in TWh from 2002 to 2017 and projected to 2030 based on estimates by the International Energy Agency. Light blue is coal, dark blue is natural gas, and green is renewables.

Back to the emissions from the Tesla. It takes about one pound of coal (0.45 kg) to generate one kilowatt hour (kwh) of electricity. So, each Tesla needs 2,250 kg of coal to produce the electricity for its yearly travel. It turns out that coal has a carbon intensity of about 1,000g CO_2/kWh, oil is 800g CO_2/kWh, natural gas is around 500g CO_2/kWh, while nuclear, hydro, wind and solar are all less than 50 g CO_2/kWh. The two metric tonnes of coal generate 5,000 kWh of electricity to power the

Tesla for 15,000 miles, which releases around **5 tCO_2e** — a little more than the petrol model. If emissions are your priority, until there are none left to buy, maybe get a second-hand petrol car.

Chinese company Broad Group erected a 10-storey apartment block in just over a day in Changsha, China.

The insatiable monster

Here is what we are told. If development stops, so does growth, and in the blinkered neoliberal view of economics, growth is essential. If we fail to grow, we are in recession, a sustained period of weak or negative growth in real GDP (output) accompanied by a significant rise in the unemployment rate. Should weak growth continue, then the economy enters a **depression**, a major downturn in the business cycle characterised by sharp and sustained declines in economic activity; high rates of unemployment, poverty, and homelessness; increased rates of personal and business bankruptcy; massive declines in stock markets; and great reductions in international trade.

All of these things are bad. And to avoid them, we are told that growth must be positive—development is essential. Having set this requirement that development is essential, it cannot stop, or all the bad things happen. We have set ourselves in a bind.

Today, we need assurances that development will not stop. This is the mainstream logic of development. It has little to do with resources but everything to do with the insatiable monster of growth. Economic growth must happen, or the economy will fail, which is a disaster for everyone.

The process of wealth creation requires infrastructure, energy, materials, and manpower—actual people or fossil fuel slaves. It also needs space, land on which activities can happen and where some resources are grown. What technocrats like to call **natural capital**—the Earth's natural resources and ecosystems that provide valuable goods and services to humanity, including air, water, soil, biodiversity, and landscapes, which are essential for life, supporting economic activities, and fostering human well-being—is converted into cash and commodities. Items that are easily used in the marketplace. This is the capital-to-cash version of development.

Eventually, this process uses up the natural capital, and development switches to alternative sources of raw materials. The 'Don't look up' asteroid on a collision course to obliterate the earth is suddenly a mighty fine ball of rare earth metals ripe for chip making. In a capitalist world, the premise is always that natural capital is sufficient to allow development to take place and deliver everything we ever wanted. The more egalitarian among us are not so cocksure and are worried that the resources will run out before everyone has a chance to develop. In the process, it will not only cause detriment to the planet but also our hopes of achieving wealth and well-being.

Everyone should assume that resources are finite and development is not inevitable because it uses up the resource base. If there is an upper limit and eventually a resource runs out. It is possible to find ways to be frugal and efficient to extend the life of the resource, and the inter-generational equity argument buys time for new resources to emerge. Wind, wave, and solar power to replace fossil fuels is a topical example, except for geothermal energy, which will be an answer for clean baseload power. But all these versions still feed the insatiable monster.

The International Renewable Energy Agency has reported that 14,438 megawatts (MW) of geothermal power was online worldwide at the end of 2020, generating 94,949 GWh of electricity.

Development stops eventually.

What about a paradigm where **development stops when everyone is developed**? If everyone has what they need for a long, healthy life full of well-being, then there is no need for further development. All we have is the replacement of worn-out goods and the delivery of ongoing services. A kind of equilibrium condition—maintenance and even improvements through innovation here and there, but not development, the *addition of components*. In theory, this is a logical position.

Full development for everyone is a long way off if maintenance only happens when everyone is a billionaire. That may be the logic of the neoliberal who would see development go on forever. But in the real world, everything is developed as far as is practical. The planet cannot support 8 billion human billionaires living a high life. Imagine the yacht moorings we would need. Indeed, who would serve drinks on the yacht?

So the stop happens somewhere less ostentatious, perhaps around universal basic needs. Every human has food, nutrition, shelter, education, safety, opportunity and a lifetime assurance that these things will persist. All very socialist, perhaps, but a necessary property of full development. When the development process has supplied these things, it can stop.

Just as my stash of one-dollar bills for breakfast declines to nothing, so too does the apple orchard have a finite lifetime, no matter how well it is managed. My breakfast does not persist even if I eat apples. Nature cannot maintain human use of natural capital indefinitely if that use erodes the natural capital. Except that is exactly what happens. So here is a solution to the breakfast conundrum. The best way to have breakfast is to **fast and not break your fast until lunchtime.** I did it, lost 10kg and feel much better.

Burning coal and oil gives us the unprecedented horsepower to alter nature, boost primary production into food, and manufacture all manner of stuff. As a resource, it will run out before more is made. However, for society and economies, coal can be replaced with other energy sources. So, most of what coal does means finding the horsepower somewhere else.

Key points

- Finite resources, such as the stash of $1 bills or coal reserves, will inevitably be depleted if continually consumed without being replenished or replaced.
- Renewable systems like an apple orchard can theoretically provide sustenance indefinitely, but only if the nutrients removed are consistently replenished without damaging the soil.
- Many current technologies and economic activities billed as green, like electric vehicles, may not actually be that colour if their inputs rely on depleting finite resources like coal.
- The prevailing paradigm of continuous economic growth and development requires perpetual extraction and depletion of finite natural resources.

In the next chapter

Our impactful 'neatness gene' influences humanity's relationship with nature. Next we take a journey through Sweden's forests and global food production challenges to understand why our instinct to tidy up landscapes might be jeopardising ecological harmony.

Chapter Six

Why are humans tidy?

Humans simplify landscapes and exponentially increase food production, which presents economic opportunities and ecological challenges. It also seems to make them want to tidy up.

What Chris says

Visiting rural properties to give advice on the management of natural areas, I am often confronted with the results of what I call the 'neatness gene'. To some degree, all humans seem to have the desire to impose order on nature. This can be seen in town as manicured lawns and in the country as large areas of one crop or as pasture consisting of one or only a few grass species. Many a time, I have arrived to look at a piece of bush that the landholder is keen to manage to be told with some pride that "I got in and tidied up a bit". To my dismay, what I see is a patch of forest that should be characterised by several layers of vegetation—forbs and grasses, shrubs, young trees with a canopy of mature trees—reduced to a park-like stand of just those mature trees.

Not only does this reflect a misunderstanding of what a natural bush is, but it also gives a glimpse of humans' primal need for a particular landscape 'look'. What humans find attractive is remarkably similar across the world. Think art, music, food, popular culture—other humans!

So it is with landscapes that city parks across the world mimic—open spaces of mown grass interspersed with trees. Often evenly spaced trees. Research shows that even humans who do not live in landscapes like this prefer these savanna-like landscapes. Why? Because this is the landscape type that our species *Homo sapiens* evolved on. This is our ancestral home. So this 'savanna look' gets unwittingly imposed on where we live. We mow, we hedge, and we plant...in neat rows.

The point is that not every habitat looks like a savannah. Most do not. Ecology often functions best in a seemingly random assemblage of many species. Like that bushland, I go out to look at. Bushland, forests, swamps and coral reefs are generally not neat - they are seemingly random collections of a myriad of life forms - they are to the human mind, messy, and our instinct is to tidy up. When we tidy up nature, we simplify ecology. This simplification leads to a decline in the diversity of organisms and the loss of ecological function. Ecological function is what delivers ecosystem services - clean water, fresh air, food and fibre - all the things that are essential for life.

We would all be better off if we let nature organise itself and resist that tidy gene. Except that for all the tidy gene is prevalent. It impacts land use, land management and the local park. It is also about control and even arrogance that humans can determine everything about the environment.

The urge to tidy is also a tendency to simplify, and this reduces the complexity of ecological systems, and this, we know, reduces their resilience. Be wary of the tidy gene.

Ecological knowledge

When a tree grows, it absorbs carbon. A moment's thought and this is obvious. Saplings are small, with few leaves and only a little wood. A mature tree has a thick trunk made up of 50% carbon, along with a canopy of branches, twigs and leaves. Belowground, the roots make up at least a third of the tree's total biomass. Carbon is captured and stored in the tree until the tree matures. Mature trees don't absorb much carbon at all. Indeed, some old-growth forests are net emitters of carbon. As dead leaves and branches decay, they emit more greenhouse gas than is absorbed by the modest amount of the new growth in the trees.

Sweden has many forests, around 30 million ha or 70% of the land area, with some 22.5 million hectares of this as productive forest land. Sweden grows more trees than it chops down, and all this sequestration makes Sweden carbon neutral.

> *The total volume of standing wood stored in Sweden's forests has more than doubled over the past century, and its forests are still sucking in a net 48 million tonnes of CO_2 a year as they grow, with another 7 million stored in long-lasting products made from Swedish wood. Taken together, that's enough to make Sweden effectively carbon neutral.*
>
> Richard Orange, Sweden's Green Dilemma: Can Cutting Down Ancient Trees Be Good for the Earth?

Timber volume grows at around 120 million forest cubic metres per year. Sweden's forestry industry harvests around 90 million forest cubic metres. However, there are other ways to report this dynamic that are less flattering. Sweden has increased the volume of timber steadily since the 1950s.

There is a dilemma, though. Validity for this green claim of carbon neutrality comes about only when the bulk of the trees in the forest estate are young. In other words, they have to be harvested after 50 years of growth and new trees planted to keep the sequestration rate high. Whilst felling trees is good for business—imagine Ikea without access to timber products—there are costs to the biodiversity of even-aged stands regularly harvested. Despite making up less than one per cent of the planet's forested land, Sweden is the world's second-biggest exporter of pulp, paper and sawn wood products. Conservationists will rightly tell you that many animals, plants and microbes only persist in old-growth forests, and these species are under pressure when their habitat declines. Replacing old-growth forests with monoculture plantations is bad for biodiversity.

Swedish forestry makes the country carbon neutral only when it accounts for the ecology and this means understanding the processes in nature. I say this glibly but realise the magnitude of this question. Ecology is complex. It captures time from evolutionary to ecological to regular day-to-day time and an appreciation of space from landscape to location. It needs some chemistry, physics and the niche sciences of geology, soil, geomorphology, meteorology and climate. It also helps to know some math, statistics and especially the scientific method.

Ecologists are the true jack of all scientific trades and masters of integration. Study ecology and you will be astonished by how ecological knowledge can help find solutions to human predicaments.

How full is the food glass?

Humanity's food production has grown exponentially over the last 100 years, reaching unprecedented levels. It will continue to grow for decades and must persist at these levels for at least another 100 years for humanity to avoid catastrophe.

Now, depending on your view of the glass, this is a half-full or half-empty situation with so many customers. There is serious money to be made, a huge economic opportunity for those who can grow food profitably or assist those who do. Plenty of people means huge amounts of brainpower and inspiration to solve whatever food production and diet problems come our way, not to mention the opportunity to build and sell technologies—**clearly a half-full glass**.

Servicing so many customers requires a huge amount of resources for the production and distribution of food, which uses up natural capital and leaves a trail of external impacts on the environment. Food supply chains are vulnerable to shocks from local to global scale, leaving an impression of fragility rather than stability and at risk of disruption from economics as well as nature—**this glass is emptying fast**.

Both options have merit. The glass of human development has been brimming over for centuries and ever faster since the capture of energy from fossil fuels. Technology helps with any number of challenges, and the dominant capitalist model of finance provides more than enough incentive for technologies to flourish. However, the requirement for 22 trillion kilocalories a day of nutritious food is daunting. There were a mere 5 million people alive at the time agriculture was invented, coincident with the last glacial maximum. Since then, it has been warming, and we now have 8 billion people, with 7 billion of them arriving in the last

200 years, most of the usable agricultural land already under production, the oceans overfished, and limits to key nutrients in sight.

So how do we choose? Half full or half empty. Here is a cautionary tale that might help your choice...

The Lily Pad Puzzle

Imagine a small pond that has clear blue water reflecting the summer sky.

In the centre of the pond are two Lily pads, the emergent leaves of an aquatic plant, floating safely on the surface of the water, curled up edges keeping the surface of the leaf dry.

On one Lily pad is a green frog.

The frog is hard to pick out on the green pad but a yellow stripe on its back gives it away.

Chance happens that a week later you pass by the same pond and stop to admire the scene. Sure enough, the frog is still there, only you notice that there are now four Lily pads; the number of leaves has doubled. In a week the frog has gained surface real estate.

A week later you happen to pass by the pond again and even before you spot the frog you see that there are now eight Lily pads. These aquatic plants are quite prolific.
You have to go away for a while and forget the frog and his growing number of Lily pads. A few months pass.
Delighted to return home you saunter by the pond again.
The first thing you see is that the pond is now mostly green as half of it is covered with Lily pads and rather less of the blue sky is reflected in the water.
Sure enough, the frog is still sitting proudly in the middle.
The puzzle question is this.
How many more weeks until the pond is completely covered with Lily pads and the frog can hop to shore without getting wet?

The problem with a half

Half is awkward for a finite resource. It is the point where, just before or just after, there is either more or less than there was before. A kind of tipping point. Before one-half, there is plenty left in the tank. After one-half, there is always less. And this is the reality for natural capital. In principle, much of what nature provides is renewable, making it hard to consume more than half. Only the current global demand is unprecedented, and we are over half of just about all the critical resources and services humans rely on.

This brings us back to Lily pads and the answer to the puzzle. How many more weeks until the pond is completely covered with Lily pads and the frog can hop to shore without getting wet? **One.**

Key points

- Humans have an innate "neatness gene" that drives us to impose simplified, savanna-like landscapes on natural habitats, but this tidying up reduces important ecological complexity and function.
- Sweden's forestry industry makes the country carbon neutral by maintaining high volumes of growing trees, but converting biodiverse old-growth forests to managed monoculture plantations involves ecological trade-offs that need to be carefully balanced.
- Humanity's exponential growth in food production represents an immense economic opportunity and an unprecedented strain on finite natural resources and ecological systems, a "glass half-full or half-empty" paradox.
- The "Lily Pad Puzzle" illustrates how exponential growth in a finite system can rapidly go from half-full to total saturation within a short doubling time. This non-linear dynamic applies to humanity's current demands on critical natural resources.

In the next chapter

The paradox of our hunger for never-ending progress against the finite limits of a planet powered by depleting fossil fuels suggests an imminent shift towards more sustainable, land-centric living as an alternative to the current growth-driven model.

How is it that in a book entitled Sustainability, this is the first use of the term six chapters into the topic? It is time we took a look at sustainability and that great oxymoron, sustainable development.

Chapter Seven

Oxymoronic

Humanity has binged on growth over the last 150 years thanks mainly to the massive energy subsidy from oil, natural gas and coal. We love growth so much we want it to go on forever.

Growth maintained?

Sustainable means maintaining a certain rate or level; to keep going. **Development** means to *create growth, progress, positive change or the addition of physical, economic, environmental, social and demographic components*, a bigger and better forward momentum.

Put these two words together, **sustainable development**, and we have ourselves an oxymoron—a figure of speech in which contradictory terms appear together. Development as economic growth through addition cannot go on indefinitely because of the relationship between economic activity and resource consumption... on a finite planet.

There is only so much food that can be grown in soil. Freshwater availability is constrained in time and location. Habitable land is limited. The fossil fuel that drives our economies will run out or pollute us out of existence. There are probably about 50 years of oil left at the current use rate of 100 million barrels a day.

Nature has a thing or two to tell us, too. There is tension between continuous growth and keeping things going because this is not what happens anywhere in the natural world. In nature, where the single primary energy source is the sun, sooner or later, growth slows, stops and what was built collapses. Admittedly, what is lost is replaced by another set of organisms mopping up the new resources, but indefinite growth is not how nature works, at least not on a finite planet.

Rainforests are home to half of all the living animal and plant species on the planet.

A rainforest looks lush and full of life growing all around. Plants and animals do grow and reproduce, but they also die. What is happening is not continuous growth but recycling. Energy and nutrients are transferred between temporary storage by myriad organisms doing their best to defy entropy.

A rainforest looks lush and full of life because this cycle is rapid. It looks stable because the trees are large and long-lived, the weather is predictable, and it takes a severe storm to knock about the multi-layered vegetation that fills all the available space and protects the soil from the sun. The capture of the necessary energy to defy entropy can only be maintained for a finite amount of time. The rainforest was not always a forest. A subtle shift in the climate, and it might not be again. A cyclone disrupts any temporary balance, as does a drought or a wildfire. And then, over geological time, not only does the climate change, but the tectonic plates move around, reconfigure and send ocean currents in different directions.

Nature copes with all this, appearing to stay stable by accepting disturbance and change. It absorbs disturbance as a critical driver of evolutionary processes, resulting in the diversity of life. There is always an alternative organism waiting for an opportunity. Nutrients and energy continue to cycle through a combination of ecological adaptation by some organisms and the evolution of others. Millions of individual organisms, from microbes to mammals, look stable in the aggregate, doing what it takes to defy entropy and reproduce.

But stability is temporary. Nature can do balance, but mostly it does change.

An inevitable phrase

When you look at the synonyms for sustainable and for development, it was inevitable that someone would put the two together to create a wonderful, inspiring phrase that would capture the essence of the modern human—grow forever.

Synonyms for sustainability include

> *stamina, supportability, resilience, vitality, achievability, stability, reliability, acceptability, practicality, autonomy, profitability, cost-effectiveness, effectiveness, equilibrium, stabilisation, perseverance, tenacity.*

Synonyms for development include

> *advancement, betterment, improvement, perfection, refinement, incubation, maturation, maturing, ripening, blossoming, flourishing, flowering, addition, augmentation, enhancement, supplementation, emergence, evolvement, metamorphosis.*

Getting better, doing better—everything better is in our genes. It drives us to take risks to make more. And we have become so good at making more that we ignore what we know about nature and the finite dimensions of space and resources on this tiny blue planet. At least, that is what we tell ourselves.

Our delusion of continuous growth has come about because we harnessed a pulse of ancient free energy in the form of fossil fuels. Without that input, we would not have modern agriculture, and we have converted billions of tons of food into more humans. Admittedly, we will probably invent another near-free energy source to replace this pulse because societies would collapse in a few days without energy. Develop or die, but even if fusion or solar panels in space or deep rock heat get us out of the fossil spike, resources to make goods and infrastructure are finite. Most will run out sooner than later.

Humans don't do balance

Development—coming into existence or creating something new or more advanced—is an attractive antidote to the threat of entropy. It has all the moving forward, resource use and defiance to keep going. It makes biological as well as human sense for the individual.

Humans have become masters of creating something new. What we call development has changed lives throughout history, mainly for the better. Sure, there are negatives in the development of weapons, inequitable economic systems, reality TV, and the like, but modern humans are materially and economically way better off than our ancestors. Why wouldn't we want to keep going?

Another retort to all the oxymoron negativity is that sustainable development is not about 'growth that uses resources' but progress toward greater well-being for everyone. This is the essence of the 2030 Agenda for Sustainable Development, adopted by all United Nations Member States in 2015...

A shared blueprint for peace and prosperity for people and the planet, now and into the future.

The agenda gives us the 17 Sustainable Development Goals (SDGs) that, if achieved, would end poverty and other deprivations through strategies that improve health and education, reduce inequality, and spur economic growth—all while tackling climate change and working to preserve our oceans and forests.

SUSTAINABLE
DEVELOPMENT
GOALS

A laudable, politically correct agenda that can only happen if there is development to generate wealth because the core assumption in the goals is that the current economic model is the only one we have to end poverty. A term defined as the state of being extremely poor. But we know that most wealth begins with nature. The first steps out of poverty are access to food, clean water, and shelter, which are nature's services. The market begins by selling fruit, vegetables and bushmeat.

Modern economies can grow without natural resources because the production and consumption of goods and services are not based on nature. Not all capital and rent require land, soil, minerals or oil, but these traditional sources of capital are still effective. But it takes a while before markets trade in derivatives. Before then, they trade food and raw materials. And when we make new people, they need 2,500 kilocalories worth of food every day, have to make a living and have somewhere to live.

Urban Growth May Not Happen As Imagined

Urban growth is trending to put 4 out of 5 people in urban areas. This steady urbanisation comes from the need to accommodate and support population growth that took off in the 20th century and added 4 billion people in a hundred years.

In 2020, when we all tried to curb the spread of COVID-19, 56% of us were locked down in cities and urban areas. In Europe, the proportion was 75%, and in Northern America, it was 84%.

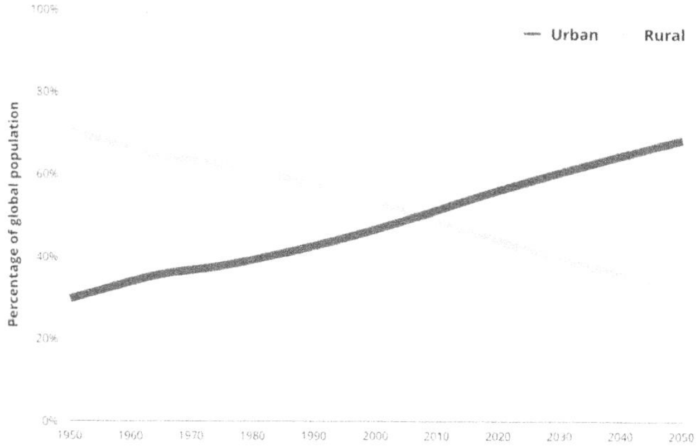

UN records for the trend in urban growth and the projection for 2050.

The explanation is that cities are where opportunities happen, whereas rural areas offer hard work for little financial reward. Cities provide the chance of riches and a more comfortable lifestyle from the trade of physical labour for wages, even if riches are nearly always elusive. Unless you can become a landowner, rural life is trading time for money at modest rates.

The thermodynamic reality is that rural areas, especially in countries with mature economies, receive an energy subsidy from fossil fuel inputs in return for producing excess food to feed the urban populations. This is the necessary energy conversion and transfer that allows people to live in large groups. In their tractors and machinery, farmers have the power equivalent of a slave army that helps them till fields, plant, nurture, harvest and store food crops and rear livestock to transport to cities. A single barrel of oil (42 US gallons or 159 litres) refined and burnt in machines can deliver power equivalent to about 1,700 kWh of work, while a healthy human labourer can perform about 0.6 kWh in one workday.

Suppose the farmer uses a barrel of oil across his farm machinery each week and in the inputs of fertiliser and pesticides. In that case, it is as though he has commandeered 1,500 workers, even if we assume that oil is less efficient at work than pairs of hands. If the UN projection for 4 out of 5 people living in large groups with limited capacity to grow their food is correct, what does it say about this virtual rural labour force?

Before finding an answer, we need to capture another assumption and figure out what the global population will be in 2050. Again, the UN tells us this figure will be 9.7 billion. Part of this increase is baked in because of the population growth in Africa and Asia that has already generated large numbers of youngsters ready to reproduce themselves. All that needs to happen is for these people to remain fed sufficiently to survive and reproduce. The likelihood of this is high given that enough food is grown globally and there are enough fossil fuel inputs available for this to continue for a decade or two. Food shortages are more about distribution and access than production. A catastrophe, even a collapse, is possible, but current global food systems could result in 9.7 billion people.

Given the urban growth trend, by 2050, there will be 6.8 billion people in towns and cities fed largely by the efforts of 2.9 billion living in rural areas. This rural population will be 500 million fewer than the current rural population. This rural community will grow 24% more food than today, supported by the virtual army of labourers provided by the power of fossil fuels or alternatives. The assumption in the population projection and the graph of where these people will live is that **the energy subsidy to food production will continue and grow by a quarter in a generation**—roughly 1% per annum. Fewer rural residents growing ever more food.

Development is at the core of the demographic transition

What if the energy subsidy fails?

The extra energy used to grow food is not a subsidy by the government, although many governments do support farmers. It is an **energy subsidy** that is a consequence of intensive agriculture. Machines and inputs (fertilizers, pesticides and water) from outside the farm are energy sources for the crop or the livestock. The energy used to power the machines and energy embodied in the extraction and manufacture of all the tools and inputs a farmer uses.

It is hard to quantify exactly how much external energy it takes to put food on a person's plate in New York or Mumbai, but here is a recent approximation from the US.

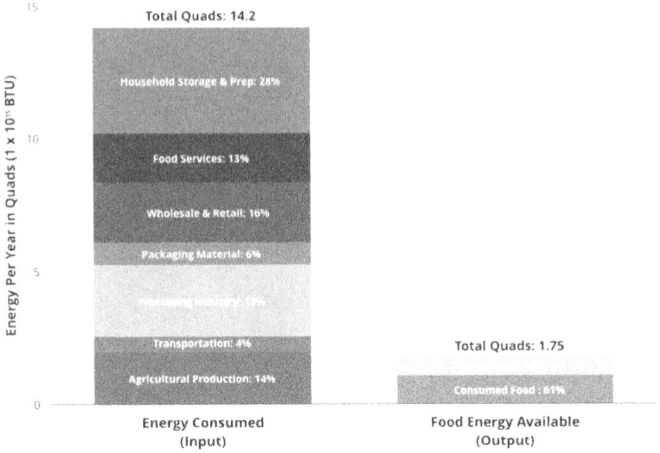

Illustration of the reliance on exogenous energy, energy from outside the body. Source: Bradford, J., 2019. The Future is Rural: Food System Adaptations to the Great Simplification.

In this analysis, roughly **eight units of energy** are needed to get one unit of energy into an urban human. If food waste is shifted to the supply side, the ratio rises to 14 to 1. Even if agricultural production is efficient, the food supply system isn't. And this is just for energy; we could run similar numbers for resources and nutrients.

At these ratios, the energy subsidy cannot fail. If it does, millions will starve, most of them in cities and towns aggregated from the consequences of urban growth. The historical famines that predominantly affected rural areas due to drought and widespread crop failures will march into the cities.

The stark reality is that the current forms of energy the farmer uses will not persist beyond 2050. Fossil fuels will either be phased out to reduce greenhouse gas emissions or become too expensive to extract and use. This will not be sudden for a couple of reasons. First, the machinery on farms is predominantly powered by internal combustion engines. Replacing these machines with electrical-powered options will take more work and resources than are currently available. Nor will replacing or retrofitting all the road transport vehicles. The ships that take the food worldwide will be especially tough to convert, probably to gas as an interim fuel or directly to hydrogen—the wind is even a possibility.

The bigger challenge is that alternatives need to be at scale. The volume of production, transport and processing is off the charts. And some can never be because limited mineral resources are available to manufacture them. Then, there are the industrial processes that produce nitrogen fertilisers. They use gas as a raw material and a lot of energy. However, producers will experience yield gaps and lower production without fertilisers, so this will likely be the last of the subsidies to end.

Energy subsidies will only continue if an alternative energy source comes online. Even then, there are limits to the supplies of raw materials to retrofit or replace all the existing farm machinery and infrastructure. In short, the future of the energy subsidy to agriculture is perilous.

People on the land

When the subsidy declines, it will mean changes to intensive agriculture. What will happen is that the alternative food production systems will be more labour-intensive, local, closed, and rely on nature to maintain yields. More people will grow food, even those still living in cities.

The solution to not having enough energy inputs on farms is getting people and animals to do the work. Farms will become more circular and self-sustaining, just as they used to be. It will be a back-to-the-future moment, with production returning to a net energy surplus. Alternatively, the whole system collapses, and we enter dystopia.

Suppose that the energy supply didn't fail but was gradually transitioned into a much lower per capita use. Transition buys time for people to adjust and for the new production systems to appear and replace the energy input-output approach we have today. What would happen is that people would return to the land. There would be more people growing a proportion of their food. Here is how we know.

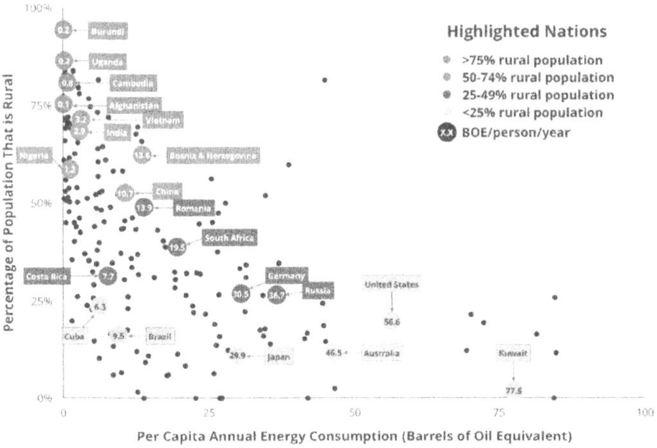

Corellation between the proportion of people living in rural areas and the per capita enrgy consumption by country.

Most countries with at least 50% of the population as rural consume less than 10 barrels of oil per person per year. All the high per capita energy-consuming countries have less than a third of their population on the land.

Returning agricultural production to a net energy source by involving more human labour in food production has bonuses. Soil health would be more easily gained and less energy-intensive if human cultures were more spread across the landscape. People would more easily eat, dispose of their waste, and maintain their animals inside the plant communities on which the entire system is based.

Instead, we have disconnected ourselves from productive landscapes, paid to clean up the pollution in cities and feedlots—cities are functionally equivalent to feedlots for people—and then paid to mine, manufacture, transport, and disperse replacement minerals back to the soil. Here is how Jason Bradford from the Post Carbon Insitute explains the luxury of long-distance transport of food to cities and long-distance transport of wastes back to farms.

> *This is called progress. It would be wise to sort out how to return to land what we take from it and dispense with energy-demanding mining and processing. Returning all human waste to fields is entirely possible and was the method by which Chinese and Japanese farmers persisted for centuries.*
>
> Jason Bradford

The Great Acceleration

Modern humans bumbled along for nearly 300,000 years, only meandering beyond Africa 70,000 years ago. However, our ancestors had phenomenal success in spreading far and wide on the back of our control of fire, communication and, eventually, technologies.

Initially, numbers did not surge. We were sparse, tribal and, at times, barely made it through. Research into the theory of Mitochondrial Eve, where all women alive now are descended from her in a direct, unbroken female line, includes nuclear DNA tests that show that, at times, the effective population number of early humans was a few thousand. The bumbling continued for 290,000 years before, in a few places, agriculture was invented, and a few recognisable civilisations appeared, flourished and fell away. It was not until the last 200 years or so of modern human existence, roughly 0.07% of our time on the planet, that the bumbling stopped and the acceleration began.

The Great Acceleration is the rapid, continuous, and approximately synchronous spike in growth rate across a wide variety of human activity measurements that began in the mid-twentieth century and continues to this day. The metrics of this growth, which can include population, economics, water usage, food production, transportation, technology, greenhouse gases, surface temperature, and natural resource usage, have all increased, often exponentially, since the late 1800s. The speed and size of this change cannot be understated.

Perhaps because of the unprecedented and, for many, desirable changes, much of the debate is about how long the acceleration will continue. Environmental historian J. R. McNeill has suggested that the Great Acceleration is unique to our time and will come to an end soon; it has never happened before and will never happen again. However, others, including some climate change scientists and chemists, suggest there is insufficient evidence to either validate or disprove such a notion.

As for the acceleration itself, it is about humans and so is seen as anthropomorphic. Here is one such summary of the cause.

> *The Great Acceleration is generated and sustained by powerful economic mechanisms, including globalization, marketization, and financialization, which are based on the mainstream model of doing business and promoting economic growth.*

Boda, Z., & Zsolnai, L. (2016). The failure of business ethics. Society and Business Review. 11, 93–104.

No, that is the ultimate consequence, **not** the proximate cause. The cause of the acceleration is the sudden availability of exogenous energy—fossil fuels—that humans converted into more humans. Sure, the mechanisms are nested in economics, but the cause is energy in excess.

Rapid, ubiquitous and copious population growth is the reason feeding everyone well is such a huge challenge. The acceleration will slow and stop even if we find an alternative exogenous energy source to substitute for the finite fossil fuels because resources are limited. How scientists can even suggest that there is "*insufficient evidence to either validate or disprove such a notion*" is incomprehensible and reprehensible. They should know better. We know you do.

It is time to stop kidding ourselves. Sustainable development is an oxymoron—a figure of speech in which apparently contradictory terms appear in conjunction. The SDGs are a grand gesture that, should they be met, are a boon to many people most in need of resources. But they are based on a false premise, implying that the growth necessary to meet the goals can somehow happen alongside preservation.

Some conservation is possible. Maybe we could rewild a third of the planet and still feed everyone well. Maybe we could all consume less, waste less and put our prejudice and fear aside. What this means is giving up growth as the paradigm that protects us from entropy. The concept that means we survive. The one we latched onto and has seen spectacular success.

Maybe we could even give up growth. We could start by trashing an oxymoron or two.

Economic growth and human population growth on the back of fossil fuel subsidies over the past 100 years have been spectacular. A miracle. No other organism has appropriated exogenous energy and converted it into more making and aggregated populations through urban growth.

We know that organisms given excess resources can grow in numbers. Bacteria and fungi do this all the time. But we also know what happens when the resource is used up, and that is what more making does; it uses resources. So the growth projection—adding the yellow and brown projections in the graph above—is naive. Any healthy sceptic would warn against such simple predictions even if the historical trend was as predictable as this one.

The coming bust may force those that remain back onto the land. There may not be another choice. Meantime we should stop thinking that urban growth is given and figure out ways to make what's coming a transition and not a collapse.

Key points

- The concept of "sustainable development" is an oxymoron, as indefinite economic growth and development are fundamentally incompatible with the planet's finite resources. The rapid growth of the past 150 years has been fueled by the unsustainable exploitation of fossil fuels.

- Current projections of increasing urbanization, with 4 out of 5 people living in cities by 2050, are based on the assumption that continued fossil fuel subsidies to agriculture will enable fewer rural farmers to feed growing urban populations. As these fossil fuels decline, this trend is likely to reverse.

- The decline of fossil fuel-based agriculture will necessitate a shift to more labour-intensive, localized, circular agricultural systems, with more people directly involved in food production. This may lead to a reversal of urbanization trends as people return to the land.

- The "Great Acceleration" of rapid economic and population growth since the Industrial Revolution is a one-time anomaly enabled by abundant fossil fuel energy. As these finite resources are depleted, the acceleration will inevitably slow. A voluntary transition away from growth-based economic paradigms may be necessary to avoid collapse.

In the next chapter

Complaints are all very well, but what actionable steps are there amidst global uncertainty? Perhaps a shift from goals and targets to understanding processes and embracing scenario-based thinking. Let's see.

Chapter Eight

Sustainability

'Sustainability' might be popular but it is a hollow, misleading myth that fails to account for the harsh realities of thermodynamics and ecological timescales, which make indefinitely maintaining resource consumption levels an impossibility

What is sustainability?

Recently, one of my golfing buddies went on a cruise with his family to New Zealand. The ship he sailed on took in the coastal towns, majestic fjords, and scenery as it made its way around the South Island. There were many people on that boat, roughly 4,000 of them paying an average of $2,400 each for the ten-day trip.

Phil had a great time. He enjoyed all the food, entertainment and the scenery. He was a bit shocked at how people behaved in the restaurants, piling up their plates with all the foods they didn't normally eat and then leaving most of what they had piled uneaten on the plate. He especially enjoyed the presentations from the crew on the geography, biology and culture of the places the ship cruised past. One point of commentary that stuck with my buddy was that the southern tip of the island was rich in minerals, but not to worry because the South Island was too rugged to exploit them. Phew.

Cruise ships carried 31.7 million passengers in 2023

Sustainability is supposed to be about '*meeting the needs of the present without compromising the ability of future generations to meet their needs*' typically under the three pillars of profit, planet and people. However, this definition of sustainability is impossible for humans. As a highly successful species, we use resources, both natural and man-made, all the time with little restraint. Currently, overshoot day is sometime in August, meaning that the global economy has been overconsuming renewable resources for nearly half a year. Our voracity for natural resources seems unstoppable.

While all this development goes on, food production and diet delivered as they must be through ecology and nature's services will need to persist at the equivalent of at least 22 trillion kilocalories per day plus adequate nutrition for at least another 100 years. Feeding everyone sounds **very much like a sustainability challenge**. A huge need that must be met. Sustainability feels like an impossibility.

Any single action might, on its own, be sustainable—always take a reusable mug to the coffee shop. But this does not mean that the supply of expertly roasted coffee beans, milk or the power to run the espresso machine is sustainable. Sustainability is more inclusive than the personal feel-good factor of taking a reusable mug for your morning latte. It has to include the resource base, the production system, and the human choices that allow the coffee shop to exist.

True sustainability, for example, for feeding everyone well, means moving towards food ecology and diet where humanity lives off the interest of natural capital rather than disrupting nature's services eroded by our success. This is a process of continuous improvement rather than a goal or a target set from the comfort of air-conditioned offices. It may not even be possible for humanity to become truly sustainable, but trying hard to achieve sustainability is essential for a healthy and safe future.

In its current formulation, sustainability should become a dirty word and will when history unveils its hypocrisy. No surprise that I don't like the word. At least not in its greenwash, 'cake and eat it' context that we so often see bandied around. But we do have to face the realities of meeting current and future needs.

Multiple land use can help sustainability.

The sustainability challenge for the individual

The first steps as individuals might be to change light bulbs, consider diet choices, install solar panels and take a reusable mug to the coffee shop. Maybe even have a chat about sustainability with fellow caffeine connoisseurs.

It will also make a difference if personal investments are made with sustainability in mind, along with big purchases and life choices. Individuals can also push up into societal systems to create pressure for change.

Equally important for individuals is to move beyond the rhetoric of sustainability to reality and not get bogged down in the bad news cycle that seems to be everywhere. Of course, there are bad things happening that are destroying resources rather than using them wisely, but every small step away from such actions is a win.

Learn to consume evidence on what is truly sustainable to make better-informed choices about consumption and lifestyle.

The sustainability challenge for society

The first step is a collective will, a need to all agree that sustainability is not just for PR machines but is a desired state for whole economies. This is not easy and cannot happen overnight, but every move matters.

Practically, this will mean alternative energy sources, more efficient agriculture, including water use, and less pressure on food supply chains to deliver profit over nutrition.

A critical part of the collective challenge is ensuring a reliable and robust evidence base that delivers general agreement on what sustainable actions must happen.

Evidence is especially important in the small steps forward that are needed — the more efficient jet fuel aircraft as a transition to a solar or hydrogen-powered airliner — and the goal-setting that can take everyone on the journey. More important are the bigger conversations about profit, the planet and people.

"*Meeting the needs of the present without compromising future generations*" begins with the food that we eat because we have to eat, and there are now so many of us. Food production, processing and consumption require much smarter use of natural resources than the current conversion of natural capital into food commodities for profit.

One of the smarts will be to think of resources together rather than in isolation. A wheat crop might be profitable and water efficient, but it is not sustainable if the production system mines the soil carbon. We must use the available science and evidence even as we add to the mountain of human knowledge. In short, get much better at the evaluation of natural capital.

Healthy food for everyone, when there are already 8 billion souls and their companion animals, will need amazing ideas from everyone. This is a prod for adult discussion on sustainability that must be had everywhere. In this guide are some prompts for the conversation.

The injection of exogenous energy in the fossil fuel pulse has allowed humans to reproduce and use resources way beyond the earth's natural capacity to capture and cycle energy and nutrients.

Conventional wisdom has sustainability, meaning ***"the ability to be maintained at a certain rate or level"***. This can translate to 'keep going' so that constancy is maintained. This idea is attractive because it implies opportunity and survival compared to the alternative of decline or collapse. Maintained also suggests it is always possible—whatever "it" might be will continue forever.

The snag is that nature still needs to get the memo. She does not stay the same ever. Instead, nature is constantly in flux through a fantastic ability to adapt and bend with whatever environmental changes happen. Nature must be flexible and change because all ecological processes are governed by basic thermodynamics.

Let's consider sustainability from these laws of nature to understand why the typical definitions of sustainability are misleading and find out why ecological time is a much better frame, especially for the global challenge of feeding everyone well.

Sustainability with thermodynamics?

Polish-American chemical engineer and physicist Tadeusz Patzek goes back to nature to find a definition of sustainability that considers basic physics.

A cyclic process is sustainable if and only if (i) It is capable of being sustained, i.e., maintained without interruption, weakening or loss of quality 'forever,' and (ii) The environment on which this process feeds and to which it expels its waste is also sustained forever.

Professor Tadeusz Patzek, Polish-American chemical engineer and physicist

Presumably, thermodynamics was not considered in the social definition of sustainability provided in 1987 by the United Nations Brundtland Commission and most commonly cited by proponents:

Meeting the needs of the present without compromising the ability of future generations to meet their own needs.

Here sustainability seems to be about avoiding the depletion of natural resources to maintain an ecological balance and has something to do with intergenerational equity—a colossal challenge thermodynamically. In crude terms, keep what you have today so future generations can use the resource. But sustain means to keep going, not to stop. Popular definitions assume that resource use is a given because humanity needs food, water and shelter. Each person has that divine right to use resources for our needs directly from our DNA. Resource use is what we do.

In practice, most sustainability responses proposed to date conserve, not abstain. Turn off the lights, change incandescent bulbs to LEDs, reduce waste, and recycle plastics, with a host of 'saving' actions. These make us feel good because we can do something ourselves but still need to tackle the longer-term challenge of resource use and thermodynamics. And so we arrive at ecological time.

The real challenge of sustainability

In a word, time.

Use without using up is acceptable in a moment. But what if that is a year, a generation, or 100 years? Since we started using crude oil in earnest, the total consumption will reach about 1.65 trillion barrels (roughly 262 trillion litres) by 2022, 11% of it in the last five years.

As I write, it is December 2022, and the estimate for global oil consumption is 97 million barrels a day, with a projected increase to 104 million per day by 2026. At the current yearly consumption of 35.4 billion barrels, humanity will use another 1.65 trillion barrels in just 47 years—1.65 trillion barrels is the volume of proven oil reserves worldwide as of 2016. This rate of energy use cannot meet a thermodynamic definition of sustainability. We all know this as we are already cooking under the waste energy.

And what else happened to all that exogenous energy from fossils? We made more people because we eat and drink fossil fuels, turning the fossil fuel pulse into a population spike. In ecological time, typically the decades to centuries when organisms interact with the environment, humans have appropriated external energy sources to inflate our population artificially and ecologically overshoot to consume more resources than there are available.

Ecological time for our species is longer than a human lifetime and much longer than a product cycle or the term of office of the elected. We tend not to think about it. We are as blind to sustainability in ecological time as we are energy blind.

Humans use roughly 97 million barrels of oil per day or 35 trillion per year.

As we have seen, the standard definition of sustainability, the one with the most traction, is *"Meeting the needs of the present without compromising the ability of future generations to meet their needs"*, which shortens to profit, planet, and people. This could be about thermodynamics, but instead, it tends to be about the broader use of resources that generate profit. But what if it was couched in energy terms?

Meeting human needs is ultimately an energy transfer. Staying alive requires energy because the organisation of cells and chemistry in the body is not thermodynamically the natural state of matter. The organisation of cells and organs takes energy to maintain the complexity. The endogenic energy from food for metabolism is more achievable with access to shelter in the cold winter.

So here is an alternative definition for sustainability with energy in mind

> *A cyclic process is sustainable if and only if (i) It is capable of being sustained, i.e., maintained without interruption, weakening or loss of quality 'forever,' and (ii) The environment on which this process feeds and to which it expels its waste is also sustained forever.*
>
> Patzek, T.W., 2004. Thermodynamics of the corn-ethanol biofuel cycle. Crit. Rev. Plant Sci. 23 (6), 519–567.

This sustainability and energy definition better reflects the truth that for a system to keep functioning, it has to be fed and cannot deplete the sources of the inputs beyond their capacity to replenish. This applies to human bodies, all other organisms and the ecosystems they inhabit. It also applies to human social systems.

Although closer to sustainability and energy than the anthropocentric profit, people and planet, this thermodynamic definition has some implications.

1. A cyclic process cannot release waste chemicals into the environment, i.e., its net waste production must be close to zero "forever", or the context is altered, and the cyclic process weakens or loses quality.

2. A sustainable cyclic process cannot release heat into the environment at a rate too high for the Earth to export to the

universe; otherwise, the environment warms.

3. A requirement to define "forever" given that in perpetuity fails to understand nature. Patzek chose a timeframe shorter than forever, 5,000 years for human civilisation, the age of Egypt's oldest surviving wood structure.

What humans do in this version of sustainability and energy turn the sun's current and ancient energy stored temporarily in plants and livestock or for longer in fossil fuels into people—as of late 2022, 8 billion chemical engines that are converting biomass energy into heat and using a whole lot more to make that experience as comfortable as they can.

Is development sustainable?

No, development is **not sustainable**, at least not the development defined by economists that *creates growth, progress, positive change or the addition of physical, economic, environmental, social and demographic components.*

We could say **it is not sustainable** because development uses up resources or because it causes externalities (the detrimental consequences of building up wealth and opportunity). Every resource that humans have ever used generates at least some waste that we dump into the environment. For example, the IUCN estimates that 14 million tons of plastic end up in the ocean every year.

We could say **it is not sustainable** because development uses resources that are finite and so eventually run out. There are planetary limits to resource use both from depletion (finite resources) and the consequences of resource use on how the planet works (externalities again).

But we say **not sustainable** because of something more fundamental. When development is successful, **it stops when everyone is developed**. Once development is reached, there is no need to keep developing; there is an endpoint. We don't need to fulfil the definition forever. Only we are not taught to think this way.

What all this says about sustainability

Sustainability is a political and social concept. In one view, it is dominion over those resources for whatever use is seen fit with little or no concern for the future. The other view has a little more contrition and an acceptance that resources might be used up and that using up natural capital is an immoral outcome for humanity or anything else sentient on the planet.

Greenwash is in the middle of this, where the politics of either end are pulled into the centre only to mask most of the truth. Claims can be made about sustainable use of resources extending into the indefinite future when the reality is that resources are mined.

The human condition is for betterment. We all strive for it thanks to our powerful genes. But we don't need resources to get better. Well-being has never been about luxury goods or holidays to the Bahamas. We need to pull sustainability out of the political frame and make it much more about whether or not the concept can help humanity persist in spite of our economic and political systems designed to move resources around. In turn, this would give us a very different view of development.

Sustainability becomes awareness-raising to a level of understanding about how the world works rather than a goal in and of itself. True sustainability would be wonderful, but the truth is that resource consumption will happen due to human wants, not as a result of any human values. Development doesn't stop when formulated around economic growth until resources are used up and the house of cards collapses. But development could stop safely if we accepted the idea of full development.

No human can live on a mountaintop with drawing resources from the valley

Rethinking Sustainability

The concept of "sustainability" has become a buzzword embraced by corporations, governments and individuals alike. However, the common definition of "meeting the needs of the present without compromising the ability of future generations to meet their own needs" is fundamentally flawed. As a species, humans are exceptionally successful resource consumers - by August each year, we have already overconsumed the renewable resources produced for the entire year. Our demand seems insatiable.

True sustainability would require living off the "interest" of natural capital, rather than eroding and disrupting nature's services through our voracity. But can 8 billion people and counting truly achieve this? The laws of thermodynamics suggest it is essentially impossible to indefinitely "sustain" our current resource use levels.

Consider food production alone - over the next century, we will need at least 22 trillion kilocalories per day to feed everyone adequately. Doing so in a truly sustainable manner that doesn't degrade soil, pollute waters, or deplete aquifers over ecological timescales seems like a pipe dream.

Personal actions like using a reusable mug may make us feel virtuous, but do little if the entire system providing our coffee fix relies on unsustainable resource extraction, processing and distribution. We have an "energy blindness" failing to see how our entire way of life is predicated on burning through finite fossil fuel reserves at an astonishing rate.

As one scholar puts it, for something to be truly sustainable: "It is capable of being maintained without interruption, weakening or loss of quality forever, and the environment on which this process feeds and expels waste is also sustained forever." Very few, if any, of our practices meet this stringent criteria.

The hard truth is that continuous economic "development" and growth are intrinsically unsustainable. By their very nature, they aim to deplete natural resources, dump externalities into the environment, and know no endpoint. Once we've fully "developed", further growth is unnecessary - yet this idea goes against centuries of ingrained economic thinking.

Sustainability has become a feel-good, greenwashing term deployed for PR rather than reflecting hard realities. A sustainable human civilization in the fullest sense may not even be possible. "Sustainability" as a word has become muddled and hypocritical.

Rather than fixating on an impossible to achieve goal, we must become deeply aware of our ecological blindspots. We are fundamentally disconnected from the cycles and timeframes that govern ecological systems and resource replenishment. Only once we shed our collective energy and resource blindness can we start making smarter choices about what to sustain, for how long, and at what cost.

This is not a reason for despair, but for clear-eyed pragmatism. We won't find solutions by deluding ourselves with appealing but hollow definitions of sustainability. With radical rethinking and deep intergenerational perspectives, we can start making difficult decisions and trade-offs to provision a healthy future for both humans and our terrestrial home. But first we must realize the futility of endless growth on a finite planet.

Key points

- The common definition of sustainability as "meeting present needs without compromising future generations" is fundamentally flawed and impossible to achieve given the realities of human resource consumption, ecological timescales, and the laws of thermodynamics.

- True sustainability would require living off nature's "interest" rather than depleting its capital, a daunting proposition for 8 billion people. Most human activities are unsustainable by the rigorous definition of being able to continue forever without producing net waste or environmental degradation.

- We have an "energy blindness" to how dependent our lifestyles are on unsustainably burning through finite resources, especially fossil fuels. Individual "green" actions like using reusable mugs are feel-good but fail to address systemic unsustainability.

- The concepts of indefinite economic growth and development are intrinsically unsustainable, as they rely on depleting finite resources and externalizing environmental costs. Sustainability should be reframed as an awareness-raising concept to drive difficult but necessary conversations about resource use on a finite planet, rather than an achievable end goal.

In the next chapter

Being a mindful sceptic means thinking in terms of processes and scenarios, not just targets. It will help us go beyond simplistic notions of sustainability to embrace a more mature, nuanced, and ultimately regenerative approach aligned with the deeper patterns of ecology.

Chapter Nine

Where to put your energy

In this crazy neo-liberal economic world, with its praise of the individual, inequities, and endless risk, it is essential to stay positive.

How to make a difference

We're always told that the best thing in life is to be a good person. Be kind, help others and do no harm. We hope you will make a difference in improving the world through good deeds.

At least, that is what the self-help manuals and gurus from all corners tell us—every small act of kindness contributes. Even in an uncertain world, every little bit helps.

Meanwhile, humanity creates an endless supply of trauma that needs a salve. Humans have outgrown the planet, risking the collapse of everything. Most aware, mindful sceptics will ask this question.

What can I do to help avoid catastrophe?

Here are a few thoughts.

A good path.

I could improve at woodwork. I will give it a go from preparing the timber, chiselling out a few ragged joints, to hours sanding down the project, only to wait another two weeks while multiple coats of varnish dry. This process takes too long for me, so shortcuts are taken, and I end up with a shoddy job.

Even if I could practise hard to learn the skills, I need to have the patience of a craftsman but do not. I can follow the process to a result, and in this, I gain something. Scattered throughout the house are various wooden items that my hand has touched. I dare not say crafted.

What I do find in this process is a near-constant feeling of gratitude. I feel grateful to the tree and to the foresters who tended it. I am warmed by the ingenuity to engineer the tools and the generations of woodworkers who have completed the trials and learnt from the errors—Youtube hosts dozens of them. Gratitude requires awareness, which is a good path—feeling grateful means you are on it.

According to the FAO global trade in wood products increased 143% to US$244 billion between 1990 and 2019

Think process

This one is not in the self-help books, but bear with me. Global uncertainty leaking into everyday lives is a symptom of a critical phase for humanity. Human numbers are vast, and our activities have bent the planet out of shape. One response to this is to set the agenda of bending it back. For example, we might set a target of saving a species from extinction. Only this is a spurious target because there is a good chance that most species we're trying to save will not make it. We fail to meet the target and set another one and end up with shifting baselines and mission creep, which is just one of the problems with target setting.

> *Shifting baseline syndrome (SBS) is a psychological and sociological phenomenon whereby each new human generation accepts as natural or normal the situation in which it was raised. With ongoing local, regional and global deterioration in the natural environment, this results in a continued lowering of people's accepted norms for these environmental conditions.*
>
> Soga, M., & Gaston, K. J. (2018). Shifting baseline syndrome: causes, consequences, and implications. Frontiers in Ecology and the Environment, 16(4), 222-230.

Shifting baselines are a huge problem when the trend is negative. As we progress, the concept of degradation becomes increasingly diluted, resulting in lower and lower standards being accepted as the norm for habitat conditions. But there is more to this. When the smelly stuff hits the fan, humanity will decide to prioritise itself over wildlife. It is what we've always done, and there is no easy way of turning that around. The best we can hope for is to shift our thinking away from targets, objectives and goals—the endpoints— and to refocus on the process itself. Ask what we need to do to improve the planet instead of continually degrading the environment and its services.

The best way to do this is to **focus on the process, not the target, objective, or baseline**. As an example, we know that taking measures to rehabilitate degraded soils can help to restore carbon levels and, in turn, improve water retention and nutrient exchange capacities. The target might be to sequester one tCO2e per hectare per annum of restored agricultural land. But we know it's unlikely that that soil will return to anything like pre-management capabilities to hold onto carbon. Taking away the vegetation and exposing the soil pushes it toward degradation. This is hard to turn around and nearly impossible to return to prior levels unless you have several thousand years to wait for the soil to recover.

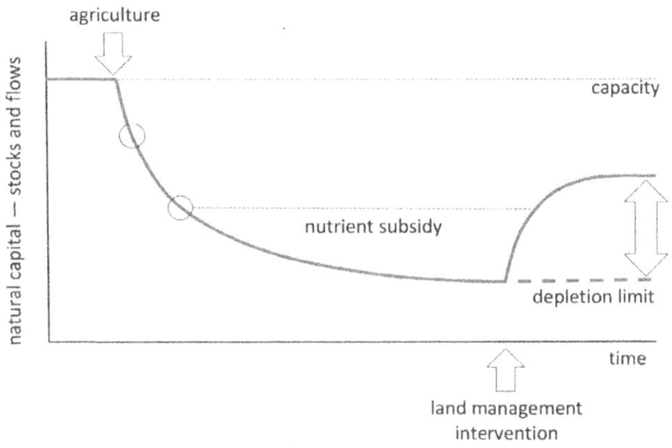

General pattern for the change in soil carbon levels after the introduction of agriculture.

Rather than the one tCO_2e ha-1 yr^{-1} target, the 'make a difference' thought is how exactly the carbon is returned to the soil. What actions need to be taken, and how do they differ from current practices? This is the point of land management intervention in the graph. The actions taken should be understood as possible directions from where we are now and move forward with awareness and understanding rather than moving forward with conventional wisdom or trying to achieve targets that can't be reached. How do we choose the actions? We go back to basics, the principles of how ecosystems function and how they work—we start with the fundamentals to know what is good ecology.

After the basics, we can ask more value-laden questions. What are the drivers of change in all directions? What are the values we want from the ecosystems that we manage?

Suppose agricultural landscapes are expected to produce food and deliver ecosystem services of clean water, clean air, pollination and a range of other regulating and provisioning services. In that case, we can move that environment towards that combination without setting a target for a particular value.

We don't need water purity to be x units, soil carbon levels to be y t ha^{-1} or air quality to z units. We need the metrics to move positively concerning their value **relative to the process**. So, objectives are set but not targets—the aim would be to maintain the value, e.g. clean air, but not a fixed amount. One thing is clear. Humanity cannot keep missing targets because that has a severe psychological effect on us as individuals and then as a collective. Missing targets has a profound impact on our mood, our collective performance, and our ability to make intelligent decisions. At the extreme, we might continue to make poor decisions because why bother?

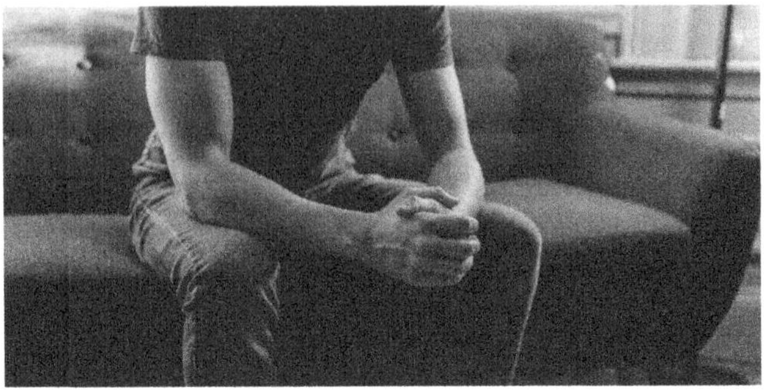

A global survey in 2021 found that more than half of people aged 16–25 felt sad, anxious or powerless, or had other negative emotions about climate change

Predict the future

Back in February 2017, I wrote a post about air travel and compared the roughly one million people in the air at any one time to a city the size of Adelaide, Australia. The COVID-19 pandemic shot a hole in that post. In a blink, the number of people able to travel by air was slashed, and projections for increases in airline passenger numbers were wrong. Nobody knew how deep the reduction in air travel would go or how long it would last. Statista estimated that it would set the global tourism industry back by $1 trillion.

This is the challenging thing about predictions. Despite the historical trend, you can never be sure what will happen next. This is the essence of Nassim Taleb's black swans. So, predictions are always tricky. Then there is the projection that you would like to see versus the projection that is reality along with a black swan around the corner—an unpredictable, rare, and high-impact event that lies outside the realm of normal expectations and conventional predictions, often causing significant consequences and altering the course of history or a system. All the uncertainty around prediction creates the perception that prediction is not a smart way to do science.

A better way to predict

A healthy sceptic has a love-hate relationship with predictions. Scepticism means being wary of predictions that assume the future will reflect the past. It rarely does, even though there is much to learn from historical trends.

Much better are predictions that reflect scenarios—situations that have a chance of happening for real. Our own Afterbefore system to predict sustainable land management is built on the comparison of possible future scenarios with business as usual. This is the best way to predict—compare options (scenarios of what could happen) with the counterfactual (what would have happened under business as usual). This is especially useful in food production. We can predict outcomes for food production with and without various interventions such as fertilizer inputs, pesticide use, changes to land management or husbandry, and even a pandemic that causes half the lorry drivers to call in sick—all sorts of changes that are within the capability of the supply chain to implement. Even those options that are currently out of reach.

We can look at the situation of what production approaches you would adopt if there were

- high-speed transport systems into remote areas
- optimal soil carbon levels
- market price halved

- fertiliser price doubled

Any number of scenarios around these kinds of questions.

We've developed the Afterbefore system to be predictive, and as we began, prediction is always only an educated guess.

We can bound the predictions with likelihoods, and we can use statistical techniques to ensure that those likelihoods have some meaning. But we are always open to not knowing the future, for this is the human condition.

In 2021 the UK Road Haulage Association estimated there was a shortage of 100,000 qualified HGV drivers across the country.

Does history help?

None of the people alive prior to the Industrial Revolution would have predicted the technology that emerged and certainly not the number of people that now live on the planet. We are aware that predicting forward is not only a challenge but comes with risk.

History is more reliable because at least some of the detail is recorded. We can piece together facts and understand how we got here at least to a point. Look backward, and we see that in the year 2000, global tourism was worth $500 billion. And as the pandemic bit value collapsed off a peak of $1.5 trillion. The growth trend over the two decades was so consistent that any projection would not have found a cliff, for there was no cliff.

Reviewing the trajectory is useful, especially if you happen to be an investor or an entrepreneur looking for an angel, but history is only useful if it can do more than show trends. It has to predict the pandemic black swan. Scenarios can reveal the black swans and, in some cases, even give them a likelihood. A pandemic was always going to happen. What was uncertain was when and how severe such a global event would be on, in this example, tourism receipts.

In 2021, the cruise industry partially recovered from 2020 pandemic lockdowns carrying 13.9 million passengers, but this was 53% lower than the pre-pandemic levels of 2019.

Being a Mindful Sceptic for the future

Make a difference by understanding ecological processes.

Again, not in the self-help books and not in the conservation manual, but it's essential. Targets are part of our psychology, but they're not helpful anymore because we act on values, not targets, even when we think they are aligned. Saving that species appears to be an admirable target; until we understand this, it is already doomed.

One of the best ways to maintain motivation and know that you are on a good path is to understand why that species is doomed.

Look at the bigger picture and the processes that allow persistence. This is about ecological thinking—more why and how than what.

So here is a way to make a difference...

- Think about processes, not targets.
- Try thinking scenarios once in a while.
- Look at all the possible outcomes, good, bad and ugly from all the possible interventions to business as usual.

And that is where to start — the 'business as usual' scenario that is sometimes called the baseline. In the case of tourism, a steady increase in global value as more and more wealth creation gives more and more people the wherewithal to travel. The project was always up because there was nothing to suggest that global wealth creation would stop or that the newly wealthy would not want to travel.

Then imagine alternatives: a doubling of wealth creation, a collapse in the global economy, virtual reality technology so good that it beats the real thing, a regional military conflict, a rogue state landing a nuclear missile...

Then think what such events would do to tourism or food production or any other of the many sustainability and environmental issues of concern.

Even a simple thought experiment is instructive. Try it—scenarios might surprise you.

Key points

- Focus on the process and journey, not just end goals. Cultivating gratitude and awareness along the way is valuable, even if the outcomes are imperfect. This applies to both personal pursuits and broader change efforts.

- Specific environmental targets, like saving a particular species, are often ineffective due to "shifting baseline syndrome" and being abandoned in crisis. It's better to focus on understanding and supporting the underlying ecological processes that enable resilience and regeneration.

- In an uncertain world, scenario planning is more useful than single-point predictions. By comparing different potential futures to a "business as usual" baseline, we can make more informed choices in the present. The COVID-19 pandemic's impact on global tourism is used as an example of an unforeseen "black swan" event that upended projections.

- Being a "mindful sceptic" means thinking in terms of processes and scenarios, not just targets. By understanding ecological principles and openly considering alternative futures, both good and bad, we can more effectively navigate uncertainty and contribute to positive change.

Science sources

All the *Mindful Sceptic Guides* are compiled from decades of experience with the science literature. Still, I also gathered specific *Evidence Reviews* that you can access on www.mindfulsceptics.info by clicking on or scanning the image below.

Mindful Sceptics

evidence reviews

scan or click the QR code for a deeper dive into the science sources for the Mindful Sceptic Guides

SCAN ME

About the Author

Mark is an ecologist, not a green one. He has built, run and sold companies, held academic posts at four universities, advised governments on environmental matters, has a PhD and still plays too much golf. His ecological research ended up in more than 70 peer-reviewed research papers and over 80 technical reports, and he supplied peer-review for more than a dozen international science journals. He lived in Africa for a decade, where he researched termites and millipedes, and once walked into a lion who snarled at him. Much of his career has been in Australia, where he won an Outstanding Teacher award from Macquarie University.

His consultancy practice advises commercial and public sector clients on assessing, monitoring, and evaluating environmental impacts, especially in carbon asset and liability management. His non-fiction books are about awareness of humans' role in nature and, as Paul Sorol, he chronicles and writes fiction for a change.

Link in bio

MINDFUL SCEPTIC GUIDES

Biodiversity loss

Escaping the Malthusian trap

Farmers

Food security

Food, people and soil

Growing enough food

Sustainable diet

Sustainability

mindfulsceptics.info

Made in the USA
Monee, IL
04 November 2026

67379699R00075